EFFECTS OF DDT ON MAN AND OTHER MAMMALS II

Papers by
Gary L. Henderson, S. M. Sieber, W. L. Heinrichs, R.W. Chadwick, Dorothy E. Woolley, Renate D. Kimbrough, Radhey L. Singhal, J.R. Brown, S. Kacew, M.C. Lechner, Janis Gabliks, E.A. Huang, P.R. Datta, Joel Bitman, Jean-Guy Alary, Joseph E. Suggs et al.

MSS Information Corporation
655 Madison Avenue, New York, N.Y. 10021

Library of Congress Cataloging in Publication Data
Main entry under title.

Effects of DDT on man and other mammals.

    Vol. 2 by Gary L. Henderson, S. M. Sieber,
W. L. Heinrichs, et al.
    1. DDT (Insecticide)--Toxicology--Addresses,
essays, lectures. I. Jukes, Thomas Hughes, 1906-
II. Henderson, Gary L.
RA1242.D35E34   632'.951   73-289
ISBN 0-8422-7110-4 (v. 1)
    0-8422-7111-2 (v. 2)

TABLE OF CONTENTS

CREDITS AND ACKNOWLEDGEMENTS

Alary, Jean-Guy; Patrick Guay; and Jules Brodeur, "Effect of Phenobarbital Pretreatment on the Metabolism of DDT in the Rat and the Bovine," *Toxicology and Applied Pharmacology*, 1971, 18:457-468.

Bitman, Joel; Helene C. Cecil; and George F. Fries, "Nonconversion of o,p'-DDT to p,p'-DDT in Rats, Sheep, Chickens, and Quail," *Science*, 1971, 174:64-66.

Brown, J.R., "The Effect of Environmental and Dietary Stress on the Concentration of 1,1-bis (4-chlorophenyl)-2,2,2-trichloroethane in Rats," *Toxicology and Applied Pharmacology*, 1970, 17:504-510.

Chadwick, R.W.; M.F. Cranmer; and A.J. Peoples, "Metabolic Alterations in the Squirrel Monkey Induced by DDT Administration and Ascorbic Acid Deficiency," *Toxicology and Applied Pharmacology*, 1971, 20:308-318.

Datta, P.R., "*In Vivo* Detoxication of p,p'-DDT via p,p'-DDE to p,p'-DDA in Rats," *Industrial Medicine*, 1970, 39:49-53.

Datta, P.R.; and M.J. Nelson, "p,p'-DDT Detoxication by Isolated Perfused Rat Liver and Kidney," *Industrial Medicine*, 1970, 39:195-198.

Gabliks, Janis; and Ellen Maltby-Askari, "The Effect of Chlorinated Hydrocarbons on Drug Metabolism in Mice," *Industrial Medicine*, 1970, 39:347-350.

Heinrichs, W.L.; R.J. Gellert; J.L. Bakke; and N.L. Lawrence, "DDT Administered to Neonatal Rats Induces Persistent Estrus Syndrome," *Science*, 1971, 173:642-643.

Henderson, Gary L.; and Dorothy E. Woolley, "Mechanisms of Neurotoxic Action of 1,1,1-trichloro-2,2-bis (*p*-chlorophenyl) ethane (DDT) in Immature and Adult Rats," *The Journal of Pharmacology and Experimental Therapeutics*, 1970, 175:60-68.

Huang, E.A.; J.Y. Lu; and R.A. Chung, "Degradation of 1,1,1-trichloro-2,2-bis (*p*-chlorophenyl) ethane by HeLa S Cells," *Biochemical Pharmacology*, 1970, 19:637-639.

Kacew, S.; R.L. Singhal; and G.M. Ling, "DDT-Induced Stimulation of Key Gluconeogenic Enzymes in Rat Kidney Cortex," *Canadian Journal of Biochemistry*, 1972, 50:225-229.

Kimbrough, Renate D.; Thomas B. Gaines; and Ralph E. Linder, "The Ultrastructure of Livers of Rats Fed DDT and Dieldrin," *Archives of Environmental Health*, 1971, 22:460-467.

Lechner, M.C.; and C.R. Pousada, "A Possible Role of Liver Microsomal Alkaline Ribonuclease in the Stimulation of Oxidative Drug Metabolism by Phenobarbital, Chlordane and Chlorophenothane (DDT)," *Biochemical Pharmacology*, 1971, 20:3021-3028.

Sieber, S.M.; and S. Fabro, "Identification of Drugs in the Preimplantation Blastocyst and in the Plasma, Uterine Secretion and Urine of the Pregnant Rabbit," *Journal of Pharmacology and Experimental Therapeutics*, 1971, 176:65-75.

Singhal, Radhey L.; J.R.E. Valadares; and Wayne S. Schwark, "Metabolic Control Mechanisms in Mammalian Systems. IX. Estrogen-like Stimulation of Uterine Enzymes by o,p'-1,1,1-trichloro-2-2-bis (*p*-chlorophenyl) ethane," *Biochemical Pharmacology*, 1970, 19:2145-2155.

Suggs, Joseph E.; Robert E. Hawk; August Curley; Elizabeth L. Boozer; and James D. McKinney, "DDT Metabolism: Oxidation of the Metabolite 2,2-bis (*p*-chlorophenyl) ethanol by Alcohol Dehydrogenase," *Science*, 1970, 168:582.

Woolley, Dorothy E., "Effects of DDT and of Drug-DDT Interactions on Electroshock Seizures in the Rat," *Toxicology and Applied Pharmacology*, 1970, 16:521-532.

Woolley, Dorothy E.; and Gloria M. Talens, "Distribution of DDT, DDD, and DDE in Tissues of Neonatal Rats and in Milk and Other Tissues of Mother Rats Chronically Exposed to DDT," *Toxicology and Applied Pharmacology*, 1971, 18:907-916.

PREFACE

This collection of recent papers (published 1970-1972) provides the latest information on the effects of DDT in humans and in other mammals. DDT is reported to display various toxic and physiological effects in humans. Among the reported effects of DDT are its influences on mammalian endocrine systems concerned with reproduction and its stimulation of various liver enzymes. Papers dealing with the degenerative metabolism of DDT in mammalian tissues and with aspects of DDT in the ecosystem are also included.

First in a multivolume collection on DDT, this volume is part of MSS' continuing series on environmental studies.

Occurrence and Physiological Effects of DDT in Other Mammals

MECHANISMS OF NEUROTOXIC ACTION OF 1,1,1-TRICHLORO-2,2-BIS(p-CHLOROPHENYL)ETHANE (DDT) IN IMMATURE AND ADULT RATS[1]

GARY L. HENDERSON AND DOROTHY E. WOOLLEY

The immature rat is much less sensitive than the adult to the lethal effects of 1,1,1-trichloro-2,2-bis (p-chlorophenyl)ethane (DDT). Lu et al. (1965) reported that the LD50 for DDT was greater than 4 g/kg the first day after birth, was 437 mg/kg at 14 days of age and decreased to 195 mg/kg during adulthood in the rat. Similarly, Henderson and Woolley (1969) reported that the LD50 was 728 mg/kg at 10 days of age and 250 mg/kg at 60 days of age in the rat. In both rats and birds symptoms of DDT toxicity are correlated with DDT concentrations in brain, and concentrations in brain are similar at death, regardless of whether the DDT is administered in a single large dose or in multiple smaller

doses (Dale et al., 1962, 1963; Stickel et al., 1966). The age difference in DDT toxicity in the rat is partly explained by the observation that central nervous system (CNS) tissues of the immature rat are less efficient than those of the adult in clearing DDT from the plasma (Henderson and Woolley, 1969). Hence, in the immature rat higher plasma levels of DDT are required to produce the same CNS levels of DDT as in the adult. However, this alone does not account for the resistance of the young rat to the effects of DDT, because after lethal doses of DDT, brain concentrations of DDT at death are higher in the immature rat than in the adult (Henderson and Woolley, 1969). This suggests that the CNS of the immature rat is actually less sensitive to the lethal effects of DDT than is the CNS of the adult.

The purpose of the present study was to analyze more closely the syndrome of DDT toxicity in the young and adult rat in order to determine whether differences in LD50's for 10- and 60-day-old rats (Henderson and Woolley, 1969) result from different thresholds for the

[1] Supported by National Institutes of Health Grant ES-00163-03.

same toxic effect or from actually different toxic actions for each age group. Death due to DDT poisoning in the adult of several mammalian species has been variously ascribed to: 1) respiratory failure because of central depression or peripheral paralysis, 2) ventricular fibrillation due to increased sensitization of the myocardium to circulating catecholamines, 3) physical debilitation because of intense tremoring and, in some cases, seizure activity and 4) extreme hyperthermia and other symptoms resulting from excessive stimulation of the sympathetic nervous system (Smith and Stohlman, 1944; Cameron and Burgess, 1945; Läuger et al., 1945a,b; Philips and Gilman, 1946; Philips et al., 1946; Judah, 1949; Hayes, 1959; Radeleff, 1964; Stavinoha and Rieger, 1966). It may be that the young rat is resistant to one or more of these effects of DDT.

Therefore, this study was designed to evaluate the relative contribution of the following to the toxicity of DDT in the adult rat: 1) respiratory depression, 2) seizures and tremoring, 3) alteration of brain function as indicated by changes in brain electrical activity, 4) cardiac arrhythmia and 5) hyperthermia. Restrained adult rats with electrodes chronically implanted in four areas of the brain (olfactory bulb, reticular formation, visual cortex and cerebellum) were given lethal doses of DDT. The effects of DDT on electroencephalogram (EEG), electrocardiogram (EKG), respiration and body temperature were recorded as toxicity progressed. Lethal doses of DDT were also administered to unrestrained 10- and 60-day-old rats, and the symptoms, frequency and severity of convulsions and changes in body temperature were compared in the two age groups.

METHODS. In all experiments purified p,p'-DDT was dissolved in cottonseed oil and administered by oral intubation. Technical DDT (Nutritional Biochemicals Corporation, Cleveland, Ohio) was purified according to the method of Cook and Cook (1946), and the recrystallized p,p'-isomer was determined to be 99.9% pure by gas chromatographic analysis. The lethal effects of DDT were studied in 10- and 60-day-old rats. All adult rats were fasted 12 hours prior to DDT administration and were provided food and water ad libitum after dosing. Adult rats were housed individually before and after DDT administration to minimize any group interaction. Room temperature in all experiments was 22°C. The mother was removed from

the young rats for four hours prior to intubation of the young with DDT. During the fasting period the infant rats were kept warm by placing a lamp over the cage. Two hours after DDT administration the mother was returned to the litter cage. All adult rats received doses of 600 mg/kg of DDT and all 10-day-old rats received doses of 1000 mg/kg. The higher dose was given to the immature rat because of the greater resistance of the young rat to the toxic effects of DDT (Henderson and Woolley, 1969). The doses administered were approximately equitoxic (LD100) to the two age groups.

In the first study, the effects of lethal doses of DDT on spontaneous brain electrical activity, EKG, respiration and body temperature were performed with five adult female Long-Evans rats (180–200 g) implanted with chronic brain electrodes. Bipolar electrodes were implanted in the olfactory bulb, reticular formation, visual cortex and cerebellum, according to procedures described previously (Woolley and Barron, 1968). Electrode positions were confirmed histologically. Spontaneous brain electrical activity was amplified with Grass P511 AC amplifiers (Grass Instruments, Quincy, Mass.) with half-amplitude high and low frequencies of 500 and 0.3 cps, respectively, and was recorded on four channels of an eight-channel Offner RC Dynograph (Beckman Instruments, Inc., Fullerton, Calif.) at paper speeds of 10, 50 and 125 mm/sec. Analysis of frequency and amplitude of wave forms was made by manual measurements of paper recordings.

Grass E2B subdermal electrodes inserted in the scapular region of the rat were used to obtain continuous EKG recordings. EKG's were amplified with a Grass P511 AC amplifier and were recorded on one channel of the Dynograph.

Respiration was continuously monitored with a Parks model 270 mercury strain gauge plethysmograph (Parks Electronics Laboratory, Beaverton, Ore.). The rats were prepared by shaving a 2-inch band around the thoracic region. The mercury strain gauge was stretched slightly around the rib cage and taped in place. As the thorax expanded, the tubing was stretched and the enclosed mercury column was lengthened and narrowed, thus increasing the electrical resistance. Changes in resistance were sensed as changes in d.c. output of the plethysmograph and were recorded on one channel of the Dynograph. Slow waves in the electrical activity of the olfactory bulbs coincided with respiration and served as a supplementary method of determining respiratory rate. This was especially useful during times of intense tremoring when movement artifacts recorded with the plethysmograph tended to obscure the respiratory pattern. Control respiration rate was determined from olfactory bulb slow-wave patterns.

11

Colonic temperature was monitored continuously with a Tele-Thermometer model 42SC (Yellow Springs Instrument Company, Yellow Springs, Ohio) with a rectal probe.

At least two control recordings of spontaneous brain electrical activity in the implanted rats were taken before acute administration of DDT. The The animals were intubated with DDT and the mercury strain gauge and EKG recording electrodes were taped in place. The thermistor probe was inserted in the rectum and then taped in place to the tail. Restraint of the animal was accomplished by taping the rat to a steel rod fixed horizontally to the recording cage floor. The rat was held in place by strips of adhesive tape placed around the rib cage and also around the tail. The bar was lowered so that the animal supported itself with all four limbs. There appeared to be no undue discomfort or interference with breathing. After DDT administration brain electrical activity, respiration, EKG and body temperature were recorded every hour at first and then continuously during the terminal stages of poisoning. Symptoms were recorded as toxicity progressed.

In the second study the occurrence and severity of seizures and changes in body temperature associated with acute DDT poisoning were determined in 19 60-day-old females and 18 10-day-old Sprague-Dawley rats. Lethal doses of DDT were administered to unrestrained rats of both ages. Symptoms and rectal temperature were recorded as toxicity progressed.

RESULTS. *Symptoms of acute DDT intoxication in 60–day-old rats.* Observations on symptoms of toxicity were carried out on 5 restrained adult rats with chronically implanted brain electrodes and on 19 unrestrained adult rats. In both groups, visible symptoms of DDT poisoning did not appear until about two hours after p.o. administration of the drug. The first symptom was an exaggerated response to sudden sound or tactile stimuli. At this time slight bursts of fine tremoring also appeared. For the next three hours the intensity of tremors increased, and there was a marked hyperkinesia. Diarrhea, salivation and a reddish secretion in the eyes were noticed. This color has been ascribed to the oversecretion of porphyrin by the Harderian glands (Carey *et al.,* 1946). Five hours after DDT, tremors were continuous and very intense. Mild clonic seizures involving the head and forepaws only also occurred, but were relatively difficult to detect because of the intense tremoring, as pointed out first by Philips and Gilman (1946). Occasional whole-body jerks

occurred about very 5 to 10 seconds. In the terminal stages, 19 of 24 animals tested showed strong bursts of seizure activity, best described as running fits, followed by motor depression. Tonic seizures were never observed. Animals which did not exhibit running fits underwent periods of very violent tremors followed by motor depression. The latter was characterized by diminution of tremors and weakness gradually progressing to flaccid paralysis in the hind limbs. The rat was unable to move and reflex movements could not be elicited in the hind limbs by toe pinch. The rate and amplitude of respiration decreased progressively. Some animals died with a terminal gasp, whereas respiration slowly ceased in others.

Two hours after DDT administration in the five adult restrained rats, the respiratory rate increased from the control level of 142 ± 20 breaths/min (mean ± S.E.M.) to 165 ± 20 breaths/min (fig. 1). This increase occurred before the gross clinical signs of DDT poisoning described above were visible. Respiratory rate reached a peak of 309 ± 30 breaths/min at three hours after DDT and declined slowly as toxicity progressed. Respiratory rate remained above control levels until shortly before death, when a precipitous drop occurred (figs. 1 and 2). This sharp decrease in respiratory rate occurred 5 to

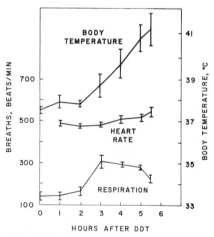

FIG. 1. Changes in respiration, heart rate and body temperature in five restrained adult rats with chronically implanted brain electrodes after a lethal dose of DDT (600 mg/kg p.o.). Each point represents the mean ± S.E.M. of five animals.

12

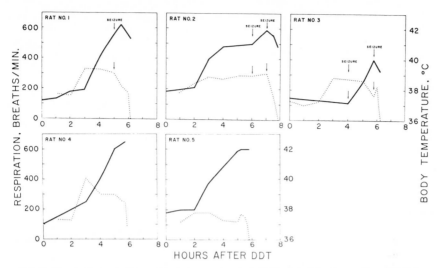

FIG. 2. Changes in respiration and body temperature in individual restrained rats after a lethal dose of DDT (600 mg/kg p.o.). Solid line represents body temperature, and dotted line represents respiratory rate. Arrow indicates occurrence of seizure activity.

50 minutes before death and during a state of general prostration.

A gradual increase in heart rate occurred during DDT poisoning (fig. 1). Heart rate was 474 ± 9 beats/min at two hours after administration of DDT and reached a peak of 544 ± 23 beats/min at 5.5 hours. No apparent abnormalities in the EKG, e.g., ectopic beats or T-wave inversion, were observed during the progression of DDT intoxication. In all instances the EKG continued for a short time after cessation of respiration and EEG.

Body (colonic) temperature remained generally unaffected for two hours after DDT administration (figs. 1 and 2). With the onset of tremoring and hyperexcitability, body temperature rose from a control level of 37.5 ± 0.2 to 38.7 ± 0.5°C by three hours and reached a peak of 41.4 ± 0.8°C at 5.5 hours after DDT. Body temperature rose as the intensity of tremors increased; in addition, a marked elevation in body temperature occurred during the motor depression which followed the intense tremoring or seizure activity (fig. 2). The range of temperatures at death was 39.5–42.5°C.

Spontaneous brain electrical activity. Spontaneous electrical activity of the olfactory bulb was characterized by slow waves synchronized with respiration and by occasional bursts of high-frequency (45–55 cps) sinusoidal waves, called *beta* waves, which were superimposed on the slower waves (fig. 3). Frequencies of the *beta* waves were similar to those reported by Woolley and Timiras (1965), who also used Long-Evans rats. Woolley and Barron (1968) reported higher *beta* frequencies (55–75 cps) for Sprague-Dawley rats and suggested that the difference was perhaps due to strain differences in the rats used.

Two to three hours after DDT administration there was a marked increase in the occurrence and a slight increase in the amplitude (25%) of the high-frequency waves in the electrical activity of the olfactory bulb (figs. 3 and 4). The frequency of the slow waves increased from about 2 cps during the control period to 3 to 5 cps at this time. These changes were coincident with the onset of symptoms of poisoning: hyperexcitability and tremors. High-frequency waves continued to increase in occurrence and amplitude as toxicity progressed. During the final stages of toxicity and after a period of intense tremoring or convulsive activity, the high-frequency waves disappeared, and there was a reduction in amplitude of the slow waves to approximately control values. As the animal

13

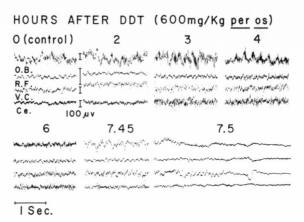

HOURS AFTER DDT (600mg/Kg per os)

0 (control)　　　2　　　　　3　　　　　4

O.B.
R.F.
V.C.
Ce.　　100 μv

6　　　　7.45　　　　　7.5

⊢——⊣
1 Sec.

Fig. 3. Sequence of changes in the spontaneous electrical activity of the olfactory bulb (O.B.), reticular formation (R.F.), visual cortex (V.C.) and cerebellum (Ce.) after a lethal dose of DDT (600 mg/kg p.o.). This animal (rat 2) exhibited seizure activity at 6 and 7 hours after DDT administration. Records at 3 and 4 hours after DDT typify increased *beta* wave activity in the olfactory bulb recording and increased amplitude of cerebellar electrical activity, whereas records at 6, 7.45 and 7.5 hours typify electrical activity during the state of prostration after seizures.

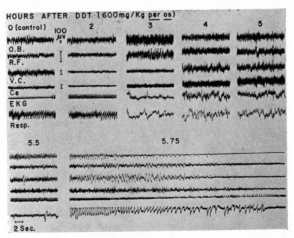

HOURS AFTER DDT (600mg/Kg per os)

0 (control)　100　　2　　　　3　　　　4　　　　5
　　　　　　　μv

O.B.
R.F.
V.C.
Ce
EKG
Resp.

5.5　　　　　　　　　5.75

⊢——⊣
2 Sec.

Fig. 4. Sequence of changes in the spontaneous electrical activity of the olfactory bulb (O.B.), reticular formation (R.F.), visual cortex (V.C.) and cerebellum (Ce.) as well as changes in EKG and respiration after a lethal dose of DDT (600 mg/kg p.o.). This rat (rat 4) did not exhibit any seizure activity, but the body temperature was over 42°C at 5.5 and 5.75 hours after DDT.

became depressed, only respiratory waves could be observed and these steadily decreased in frequency until death (figs. 3 and 4).

Amplitude of the very high-frequency cerebellar waves increased steadily as toxicity progressed, as also described by Woolley and Barron (1968) and Woolley (1970). A 2-fold increase was observed at three hours, a 3-fold increase at four hours and a 4-fold increase at five hours.

Both slow (4–6 cps) and fast (40–50 cps) frequencies were observed in reticular formation

14

electrical activity during the control period. As the symptoms of DDT intoxication became visible a 50% increase in amplitude was observed with fast frequencies predominating. During the period of depression after seizures, regular slow waves with a frequency of 6 to 8 cps, with amplitudes similar to control values, were observed.

Changes in the amplitude and frequency of electrical activity in the visual cortex after DDT administration were generally less marked than changes in other brain areas. However, in two of the five rats, the terminal stages of toxicity were characterized by a disappearance of electrical activity in the visual cortex before respiratory depression or loss of electrical activity in other brain areas occurred. In these rats, electrical activity disappeared first in the visual cortex and then in the olfactory bulb; this was followed by a decrease in respiration. As respiration failed, cerebellar and reticular formation EEG activity disappeared. EKG activity continued for a short time after respiratory failure.

In contrast, the sequence of events at death in the other three rats was, first, a slowing of respiratory movements followed by a disappearance of electrical activity in all four brain areas simultaneously. Again, EKG activity continued a short time after the cessation of respiration and EEG activity.

Hypersynchronous seizure waves during seizure activity were not observed in any of the brain areas studied.

Symptoms of DDT intoxication in the 10-day-old rat. Symptoms of DDT toxicity in the young rat began two to three hours after DDT administration when hyperkinesia and slight tremoring were observed. As toxicity progressed, hyperexcitability, tremoring and a continuous aimless wandering were noted. Tonic or tonic-clonic seizures were not observed but violent bursts of whole-body tremors occurred throughout the course of poisoning. In addition, during the latter stages of toxicity scurrying movements around the cage accompanied by vocalization were noticed. The time course of DDT poisoning was considerably longer in the young rat than in the adult. Death occurred between 10 and 14 hours in the young rat as compared with 5 and 8 hours in the adult. The terminal stage of DDT poisoning in the 10-day-old rat was characterized by a slowly progressing depression; motor activity decreased until all spontaneous

and respiratory movements ceased. At this time, movement and respiration could be elicited again if the young rats were handled. At death the pink color of the skin disappeared; the skin looked pale, was cold to the touch and no motor activity or respiration could be observed or elicited.

Hyperthermia in adult vs. young rats. Control rectal temperatures in the 19 adult unrestrained rats of the second study were $36.9 \pm 0.2°C$ (fig. 5). The control temperatures of the unrestained rats were slightly lower (about 0.6°C) than the control temperatures of the five restrained rats of the first study (compare figs. 1 and 5), reflecting the normal variability of this endpoint, probably due to differences in handling. After DDT administration, body temperature increased only very slightly in the adult rat for the first three hours and then rose sharply to reach a peak at seven hours (fig. 5). Peak body temperatures usually occurred within the hour preceding death. The mean temperature at death was $40.9 \pm 0.3°C$.

Control rectal temperature for the 10-day-old rat was $35.1 \pm 0.4°C$. This temperature corresponds with the 35°C "nest" temperature reported by Schaefer (1968) for 14-day-old rats. Body temperature decreased steadily in the young rats after DDT administration. The mean body temperature at death was $31.1 \pm 0.8°C$.

DISCUSSION. The symptomatology of DDT poisoning in both the immature and adult rat suggests that death may be due to a direct depressant effect of DDT on respiratory mechanisms at the peripheral, spinal or brain stem level. In the terminal stage of DDT toxicity in the adult rat, death was immediately preceded by a precipitous drop in respiratory rate. It appeared that respiratory failure was preceded by a flaccid paralysis of the hind limbs which steadily progressed in a caudal-rostral direction. In the 10-day-old rat, death was also preceded by a state of prostration in which all spontaneous motor and respiratory movements stopped. However, the time course of DDT poisoning in the young rat was considerably prolonged as compared with the adult. The ability of the young rat to endure the long period of depressed respiratory activity before expiring may be related to the immature rat's tolerance to anoxia, which decreases with age to reach adult levels at approximately 17 days of age (Fazèkas *et al.*, 1941).

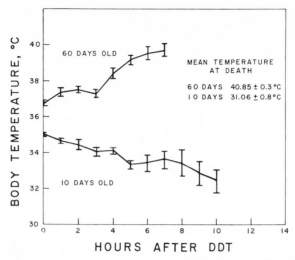

Fɪɢ. 5. Changes in body (rectal) temperature observed during the course of acute DDT poisoning in unrestrained rats. Each point represents the mean ± S.E.M. of 18 immature and 19 adult animals.

Adolph (1948) reported that the survival time for rats in an oxygen-free atmosphere was increased if the body temperature was lowered. In this connection it should be noted that the body temperature of the infant rat decreased during the course of DDT poisoning. This probably further increased its tolerance to anoxia and may help explain the longer survival time and decreased sensitivity of the DDT-poisoned immature rat.

In the adult rat, the state of prostration was preceded by seizures or marked hyperthermia, or both. In contrast, the immature rat did not exhibit seizures, and a steady decrease in body temperature was observed during the course of DDT poisoning. It is well known that the rat is not able to regulate its body temperature until after about 18 days of age. Ontogenesis of thermoregulation in the rat has been correlated with the appearance of myelin in the hypothalamus (Buchanan and Hill, 1947). Lack of fur and a high ratio of surface area to body volume permit rapid transfer of heat between the neonatal rat and its environment. Thus, as the motor activity of the immature rat increases during the course of DDT poisoning, it gets separated from the dam and its body temperature drops. The hyperthermia produced by DDT in the adult rat agrees with the hyperthermia in DDT-

poisoned domestic animals reported by Radeleff (1964). Hyperthermia may result from the intense muscular activity, from a direct interference by DDT with the temperature regulatory centers of the hypothalamus, from marked sympathetic stimulation, or from a combination of these. In mammals generally, body temperatures in excess of 40°C represent a threat to survival, and temperatures of 42°C are not compatible with survival (Belding, 1967). Therefore, in the adult rat hyperthermia is an important factor contributing to death in DDT poisoning. This hypothesis is supported by work in progress (D. E. Woolley and G. L. Henderson), in which the survival of adult rats was prolonged by preventing the hyperthermia during DDT poisoning, either by placing the animals in a cold (4°C) room or by injections of the antipyretic agent, aminopyrine.

DDT is considered to be a convulsant agent in most species (Hayes, 1959). In rats, investigators have reported that DDT-induced convulsions were either: not observed (Smith and Stohlman, 1944), rare (Cameron and Burgess, 1945; Philips and Gilman, 1946), usually observed (Dale *et al.*, 1963) or dependent upon route of administration (Judah, 1949). Some of the confusion concerning the convulsant actions of DDT probably stems from the following. 1) The mild clonic

16

movements of the forepaw and head and the whole body jerks are difficult to differentiate from the vigorous tremoring activity occurring simultaneously. 2) The more striking and intense type of seizure, which involves tonic flexion or extension of the limbs, does not occur. 3) The strong bouts of "running fits" occur terminally in the majority, but not all, of the animals.

Decreased lethal effects of convulsant drugs in infant animals are generally attributed to the fact that immature animals exhibit less severe seizures in response to convulsant agents, and so the respiratory failure associated with intense seizures does not occur. It has been suggested that this inability of the immature animal to exhibit intense seizure activity is due to incomplete development of the neural mechanisms or the specific biochemical pathways acted upon by the convulsant agents (Pylkkö and Woodbury, 1961). DDT-induced seizures are not the cause of the respiratory depression, because the latter occurs in both mature and immature rats, whereas seizures occur only in adults. Also, the convulsive episodes are not of the tonic type and so do not produce respiratory arrest by sustained contraction of the diaphragm, and the terminal running fits do not always occur. However, the motor activity of the seizures in the adult probably has a debilitating effect and in this way contributes to the greater lethality of DDT in the mature rat.

Marked sympathetic stimulation appears to be important in the symptomatology of DDT poisoning. The hyperthermia and the early increase in respiratory rate noted in this study, the EEG arousal (Woolley and Barron, 1968; Woolley, 1970), the increased urinary norepinephrine excretion (Stavinoha and Rieger, 1966), the mobilization of liver glycogen and hyperglycemia (Läuger et al., 1945a) and instances of cardiac arrhythmias in some species (Philips and Gilman, 1946) all point to a strong sympathetic involvement in DDT poisoning and should be considered in initiating therapy.

In conclusion, it is suggested that in the adult rat there are additional toxic responses, i.e., seizures and especially hyperthermia, which are not observed in the young rat and which help explain the greater lethality of DDT in the mature animal. Also, the immature rat is probably more resistant to the hypoxia resulting from the respiratory depression than is the adult.

The question of whether an age difference in DDT toxicity also exists in man may be of significance from the point of view of public health problems. The present findings suggest that if the neural mechanisms acted upon by DDT, such as those involved in thermoregulation and sympathetic responses, are not developed, then DDT will be less toxic. It is reasonable to expect that in species such as the rat, which are born with a very immature nervous system, DDT would be less toxic to the newborn than in species like the guinea pig, in which the nervous system is more highly developed at birth. The human infant is born with a relatively immature nervous system, but there is no evidence indicating the degree to which he would be resistant to the respiratory depressant and hyperthermic effects of DDT.

REFERENCES

ADOLPH, E. F.: Tolerance to cold and anoxia in infant rats. Amer. J. Physiol. **155**: 366–377, 1948.

BELDING, H. S.: Resistance to heat in man and other homeothermic animals. In Thermobiology, ed. by A. H. Rose, pp. 479–510, Academic Press, New York, 1967.

BUCHANAN, A. R. AND HILL, R. M.: Temperature regulation in albino rats correlated with determinations of myelin density in the hypothalamus. Proc. Soc. Exp. Biol. Med. **66**: 602–608, 1947.

CAMERON, G. R. AND BURGESS, F.: The toxicity of 2,2-bis (p-chlorophenyl) 1,1,1-trichloroethane (D.D.T.). Brit. Med. J. **1**: 865–871, 1945.

CAREY, E. J., DOWNER, E. M., TOOMEY, F. B. AND HAUSHALTER, E.: Morphologic effects of DDT on nerve endings, neurosomes, and fiber types in voluntary muscles. Proc. Soc. Exp. Biol. Med. **62**: 76–83, 1946.

COOK, K. H. AND COOK, W. A.: A simple purification procedure for DDT. J. Amer. Chem. Soc. **68**: 1663–1664, 1946.

DALE, W. E., GAINES, T. B. AND HAYES, W. J., JR.: Storage and excretion of DDT in starved rats. Toxicol. Appl. Pharmacol. **4**: 89–106, 1962.

DALE, W. E., GAINES, T. B., HAYES, W. J., JR. AND PEARCE, G. W.: Poisoning by DDT: Relation between clinical signs and concentration in rat brain. Science (Washington) **142**: 1474–1476, 1963.

FAZÈKAS, J. F., ALEXANDER, F. A. D. AND HIMWICH, H. E.: Tolerance of the newborn to anoxia. Amer. J. Physiol. **134**: 281–287, 1941.

HAYES, W. J., JR.: Pharmacology and toxicology of DDT. In DDT, the Insecticide Dichlorodiphenyltrichloroethane and its Significance, ed. by P. Müller, vol. II, Birkhäuser Verlag, Basel, Switzerland, 1959.

HENDERSON, G. L. AND WOOLLEY, D. E.: Studies on the relative insensitivity of the immature rat to the neurotoxic effects of DDT. J. Pharmacol. Exp. Ther. **170**: 173–180, 1969.

JUDAH, J. D.: Studies on the metabolism and mode of action of DDT. Brit. J. Pharmacol. **4**: 120–131, 1949.

17

Lu, F. C., Jessup, D. C. and Levallée, A.: Toxicity of pesticides in young versus adult rats. Food Cosmet. Toxicol. **3**: 591–596, 1965.

Läuger, P., Pulver, R. and Montigel, C.: Ueber die Wirkungsweise von 4,4'-dichlordiphenyltrichlormethylmethan (DDT-Geigy) im Warmblüterorganismus. Experientia (Basel) **1**: 120–121, 1945a.

Läuger, P., Pulver, R. and Montigel, C.: Ueber die Wirkungsweise von 4,4'-dichlordiphenyltrichlormethylmethan (DDT-Geigy) im Warmblüterorganismus. Helv. Phys. Acta **3**: 405–415, 1945b.

Philips, F. S. and Gilman, A.: Studies on the pharmacology of DDT. I. The acute toxicity of DDT following intravenous injection in mammals with observations on the treatment of acute DDT poisoning. J. Pharmacol. Exp. Ther. **86**: 213–221, 1946.

Philips, F. S., Gilman, A. and Crescitelli, F. N.: Studies on the pharmacology of DDT (2,2,bis-parachlorphenyl)-1,1,1 trichloroethane). II. The sensitization of the myocardium to sympathetic stimulation during acute DDT intoxication. J. Pharmacol. Exp. Ther. **86**: 222–228, 1946.

Pylkkö, O. O. and Woodbury, D. M.: The effect of maturation on chemically-induced seizures in rats. J. Pharmacol. Exp. Ther. **131**: 185–190, 1961.

Radeleff, R. D.: Chlorinated hydrocarbon pesticides. *In* Veterinary Toxicology, pp. 212–230, Lea & Febiger, Philadelphia, 1964.

Schaefer, T., Jr.: Some methodological implications of the research on "early handling" in the rat. *In* Early Experience and Behavior, ed. by G. Newton and S. Levine, pp. 1–40, Charles C Thomas, Publisher, Springfield, Ill., 1968.

Smith, M. I. and Stohlman, E. F.: The pharmacologic action of 2,2-bis(p-chlorophenyl)-1,1,1-trichloroethane and its estimation in the tissues and body fluids. U.S. Public Health Reports **59**: 984–993, 1944.

Stavinoha, W. B. and Rieger, J. A., Jr.: Effect of DDT on the urinary excretion of epinephrine and norepinephrine by rats. Toxicol. Appl. Pharmacol. **8**: 365–368, 1966.

Stickel, L. F., Stickel, W. H. and Christensen, R.: Residues of DDT in brains and bodies of birds that died on dosage and in survivors. Science (Washington) **151**: 1549–1551, 1966.

Woolley, D. E.: Effects of DDT on the nervous system of the rat. *In* The Biological Impact of Pesticides in the Environment, ed. by J. W. Gillett, Environmental Sciences Series No. 1, Oregon State University Press, Corvallis, in press, 1970.

Woolley, D. E. and Barron, B. A.: Effects of DDT on brain electrical activity in awake, unrestrained rats. Toxicol. Appl. Pharmacol. **12**: 440–454, 1968.

Woolley, D. E. and Timiras, P. S.: Prepyriform electrical activity in the rat during high altitude exposure. Electroencephalogr. Clin. Neurophysiol. **18**: 680–690, 1965.

18

IDENTIFICATION OF DRUGS IN THE PREIMPLANTATION BLASTOCYST AND IN THE PLASMA, UTERINE SECRETION AND URINE OF THE PREGNANT RABBIT

S. M. SIEBER AND S. FABRO

It is well known that drugs and other chemicals cross the placenta readily and reach the developing fetus; however, the mechanisms involved in this transfer process are ill-defined (Goldstein et al., 1968). It has been suggested that foreign substances cross the placenta by simple diffusion and as though the boundary had the characteristics of a lipoid barrier. Other factors such as molecular weight and degree of ionization are usually of secondary importance (Baker, 1960; Moya and Thorndike, 1962; Villee, 1965). Moreover, the morphology of the placenta undergoes changes during the different phases of gestation, and there are indications that these changes are accompanied by changes in placental permeability (Huggett and Hammond, 1952).

Most studies have focused on the transfer of drugs across the placenta during late pregnancy. Very little is known about placental permeability during the early stages of gestation (Goldstein and Warren, 1962; Brambell, 1966; Keberle et al., 1966; Fabro et al., 1967).

This article concerns itself with the transfer of some commonly used drugs from the maternal circulation into the uterine secretion as well as their penetration into the preimplantation blastocyst of the rabbit. Preliminary notes on this work have been previously reported (Sieber and Fabro, 1968; Fabro and Sieber, 1969).

METHODS. The following radioactive compounds were purchased: 1-methyl-C^{14}-caffeine (5.0 mc/mmol), 1,1-bis-p-chlorophenyl-C^{14}-2,2,2-trichloroethane (p,p'-DDT; 2.73 mc/mmol), carboxy-C^{14}-salicylic acid (1.5 mc/mmol), carboxyl-C^{14}-dextran, mw 16,000 to 19,000 (2.49 mc/g), carboxyl-C^{14}-dextran, mw 60,000 to 90,000 (0.54 mc/g), N-methyl-C^{14}-antipyrine (0.5 mc/mmol) and methyl-C^{14}-hexamethonium (1.55 mc/mmol) from New England Nuclear Corporation (Boston, Mass.); carbonyl-C^{14}-isonicotinic acid hydrazide (9.8 mc/mmol) and G-H^3-nicotine (210 mc/mmol) from Nuclear-Chicago Corporation (Des Plaines, Ill.); 2-C^{14}-barbital (1.10 mc/mmol) and 2-C^{14}-thiopental (5.19 mc/mmol) from Tracerlab (Waltham, Mass.); carboxyl-C^{14}-benzoic acid (5.75 mc/mmol) and 2-C^{14}-uric acid (8.0 mc/mmol) from Calbiochem (Los Angeles, Calif.). The purity of these compounds was checked by paper chromatography. All nonradioactive materials were of analytical purity and were obtained commercially with the exception of N-isonicotinylglycine (m.p. 229–231°C), which was synthesized according to the method of Gardner et al. (1954), and 1-acetyl-2-isonicotinyl-hydrazide (m.p. 193–195°C) which was synthesized following the method of Yale et al. (1953). Dimethyluric acid was a gift of Dr. J. M. Venditti, Cancer Chemotherapy National Service Center (Bethesda, Md.); (—)-cotinine and (—)-demethylcotinine were gifts of Dr. H. McKennis, Medical College of Virginia, (Richmond, Va.).

Animal experiments. New Zealand White does (3–4 kg b. wt.) were purchased from M. V. Zartman (Douglasville, Pa.). They were mated with bucks of the same breed at the breeding center and usually sent to our animal department the following day. The time of mating was considered hour zero of pregnancy. The pregnant animals were allowed free access to food (Purina Lab Chow no. 41) and water and were treated at 139 to 159 hours of pregnancy. 1-Methyl-C^{14}-caffeine (3.5 mg/kg; 5.0 μc/kg), carbonyl-C^{14}-isoniazid (13 mg/kg; 5.0 μc/kg) and 2-C^{14}-barbital (30–100 mg/kg; 8–10 μc/kg) were administered by gavage in approximately 5 ml of water. G-H^3-nicotine (50 μg/kg; 60 μc/kg) and 2-C^{14}-thiopental (7–20 mg/kg; 7–10 μc/kg) were dissolved in 1 ml of physiologic saline and administered by i.v. injection. Phenyl-C^{14}-DDT was administered both by gavage (700 μg/kg; 5 μc/kg) and dissolved in 0.3 ml of ethanol, by i.v. injection (400 μg/kg; 3.3 μc/kg).

Blood samples (0.5–2.0 ml) were collected from an ear vein of the appropriate animals at 0.25, 0.5, 0.75, 1, 2, 4 and 6 hours after dosing, into heparin-treated test tubes, and the plasma was obtained by centrifuging the blood at 800 × g for 10 minutes. The treated animals were stunned by a blow on the head 6 hours after dosing, except for those treated with nicotine and thiopental, which were killed one hour after treatment, or some of those given C^{14}-DDT and killed 24 hours after dosing. Animals were killed by exsanguination, and 10 to 25 ml of blood were collected from the jugular vein into heparin-treated centrifuge tubes.

The uterus was rapidly exposed and opened, and the free blastocysts, weighing 20 to 60 mg each, were removed. Samples of uterine secretions were obtained according to the method of Krishnan and Daniel (1967). Samples of endometrium were obtained by scraping the walls of the open uterus with a scalpel. Urine from the bladder was withdrawn with a syringe and added to that excreted during the experiments.

Measurement of radioactivity was carried out by dispersing blastocysts (3–6), plasma (0.1–0.5 ml), uterine secretions (10–90 mg), endometrium (0.2–0.6 g) and urine (0.1–0.2 ml) in sufficient liquid scintillation fluid in glass counting vials to give a final volume of approximately 20 ml. The scintillation fluid consisted of a dioxane–ethylene glycol–methanol mixture (44:1:5 by volume) containing naphthalene (6%), 2,5-diphenyloxazole (0.4%), 1,4-bis(5-phenyloxazole-2-yl)benzene (0.02%) and thixotrophic gel powder (Cab-O-Sil, 5%). Radioactivity was measured in a Nuclear-Chicago liquid scintillation spectrometer, model 725 or Mark I after the samples were left in the counter for 24 hours. Counting efficiency was determined by the twin channel ratio method (Bush, 1963) or by the external standard method (Wang and Willis, 1965). pH values were measured electrometrically by using a Corning model 12 pH meter.

Identification of radioactive compounds by inverse radioisotope dilution analysis. The C^{14} or H^3 compounds and their radioactive metabolites were identified by an inverse radioisotope dilution technique after solvent extraction and chromatographic separation. Tissue extracts were chromatographed with the appropriate reference compounds after their total radioactive content was determined by scintillation counting. The radioactive compounds were identified by comparing their R_f values and color reactions with those of authentic samples. For estimation of C^{14} or H^3, areas of the paper chromatograms, known to contain specific compounds, were cut out and transferred to counting vials containing 20 ml of scintillation fluid (toluene containing 2,5-diphenyloxazole, 0.5%;

and 1,4-bis(5-phenyloxazole-2-yl)benzene, (0.03%). The rest of the chromatogram was similarly cut into pieces of about 3 cm × 3 cm and counted. The total C^{14} or H^3 was then estimated, and, after correction for background, the C^{14} or H^3 associated with each compound was calculated as a percentage of the total. All of the administered radioactive compounds were subjected to this procedure, and the chromatographic recoveries were found to range between 89% and 98%. The absolute amount of each radioactive compound thus identified in tissue was calculated from its specific activity.

Caffeine. Samples of plasma (5–10 ml) or urine (10–25 ml) were mixed with equal volumes of 0.1 M sodium phosphate buffer, pH 7.4, saturated with NaCl. Pooled blastocysts (6–12) were homogenized in 3 ml of 0.1 M sodium phosphate buffer, pH 7.4, saturated with NaCl. The buffered tissue preparations were first extracted three times with 10 volumes of benzene containing 1.0 mg of carrier caffeine. The volume of the pooled benzene extracts was reduced to a small known volume (3–5 ml) under vacuum in a rotary evaporator. The water phase was then extracted three times with 10 volumes of chloroform to which had been added 1.0 mg each of 1-methylxanthine and 1,3-dimethylxanthine and 300 μg of 1,7-dimethylxanthine. The volume of the three pooled chloroform extracts was reduced under vacuum in a rotary evaporator to 4 to 7 ml. The aqueous phase was then acidified to about pH 1 with 2 N HCl and extracted three times with 10 volumes of isobutanol to which 1.0 mg of carrier 1,3-dimethyluric acid had been added. The volume of the three pooled isobutanol extracts was reduced under vacuum to 7 to 9 ml.

Samples of the remaining water phase, as well as the benzene, chloroform and isobutanol concentrated extracts, were counted for radioactivity. The radioactive compounds were then identified in each of these four fractions by paper chromatography according to the methods of Markham and Smith (1949).

Caffeine was entirely removed by the benzene extraction and represented the only radioactive compound in that fraction. 1,3-Dimethylxanthine, 1,7-dimethylxanthine and 1-methylxanthine were identified in the chloroform fractions. The isobutanol phase contained a number of radioactive spots, one of which was identified as 1,3-dimethyluric acid. The radioactivity of the water phase could not be related to any of the above methylated derivatives of caffeine; it is conceivable that it derives from the demethylation of caffeine at position one.

Nicotine. Samples of plasma (5–10 ml) and urine (5–10 ml) were brought to pH 10 with 3 N NaOH and diluted with distilled water to approxi-

mately 20 ml. Pooled blastocysts (6–12) were homogenized in 5 ml of distilled water and brought to pH 10 with 3 N NaOH; filter paper strips containing uterine secretions (30–40 mg) were immersed in 20 ml of 0.01 N NaOH. The alkalinized tissues, to which 500 μg each of nicotine, cotinine and demethylcotinine had been added, were extracted continuously for 24 hours with chloroform in a continuous extraction apparatus. The chloroform extract was reduced to a small known volume (3–5 ml) under negative pressure in a rotary evaporator. The radioactive compounds were identified by paper chromatography with the solvent systems described by McKennis *et al.* (1964).

Barbital. Samples of plasma (5–10 ml), urine (5–10 ml) and uterine secretions (30–40 mg) were brought to pH 2 with 2 N HCl. Blastocysts (7–14) were homogenized with distilled water (10 ml), and the homogenate was brought to pH 2 with 2 N HCl. After the addition of 1.0 mg of carrier barbital to the acidified tissue preparations, they were extracted four times with 10 volumes of diethyl ether. The ether extract was reduced to a small known volume (3–7 ml) under negative pressure in a rotary evaporator. Barbital was identified in aliquots of the extracts by paper chromatography with the system described by Algeri and Walker (1952).

Thiopental. Thiopental was determined in the same way as barbital except that chromatography was carried out with the solvent systems described by Cooper and Brodie (1957).

Isoniazid. Samples of plasma (20 ml), urine (15 ml) and blastocysts (6–8) from animals treated with carbonyl-C^{14}-isoniazid were treated with 4 volumes of a mixture of 95% ethanol–acetone (1:1, v/v) after the addition of 1.0 mg each of carrier isoniazid, acetylisoniazid, isonicotinuric acid and isonicotinic acid. The samples were left at +2–0°C for about 2 hours and, after centrifugation (15 minutes at 900 × g), the supernatant was reduced to a small known volume (2–5 ml) under negative pressure in a rotary evaporator. Aliquots of the supernatant were chromatographed with the solvent systems described by Zamboni and Defranceschi (1954).

Experiments in vitro. Fourteen to 20 blastocysts obtained from six-day pregnant New Zealand White rabbits were pooled and incubated in 10 ml of Ringer–phosphate buffer containing the appropriate radioactive compound at a final concentration of 1.5 × 10^{-3} M. The radioactivity in the medium was approximately 0.05 μc/ml. The Ringer–phosphate, pH 7.2, was prepared according to Davson and Eggleton (1962). Incubations were carried out at 37°C in air atmosphere with an Aminco constant temperature water bath. At intervals, usually 5, 10, 15, 30, 45, 60, 90 and 120 minutes after the incubation had begun, two or more blastocysts were removed from the medium, blotted dry, weighed and placed in glass scintillation vials for analysis of their C^{14} content. Duplicate samples (0.1 ml) of the incubation medium were also removed at the same time, and the radioactivity was estimated. The concentration of the nonradioactive compounds, sulfanilamide and p-aminobenzoic acid, was measured in the blastocysts and the medium by the method of Bratton and Marshall (1939).

The penetration of radioactive compounds into the rabbit blastocyst was expressed on the basis of the relative amount of radioactivity in the blastocyst as compared to that in the incubation medium, that is, by the ratio C^{14} in the blastocyst (disintegrations per minute per gram)/C^{14} in the medium (disintegrations per minute per milliliter). Equilibrium is attained when an increase in the time of incubation does not result in a concomitant increase in this ratio. Half-equilibrium time ($T_{1/2}$ eq) has been used as a measure of the rate of penetration of radioactive compounds into the blastocyst. $T_{1/2}$ is defined as "the time (in minutes) which is required for a compound, incubated under the conditions described above, to reach a concentration in the blastocyst equal to one-half of that at equilibrium." $T_{1/2}$ eq was obtained by calculating a regression line with the reciprocal values of the ratio disintegrations per minute per gram of blastocyst/disintegrations per minute per milliliter of medium and the corresponding time of incubation (in minutes). By this procedure, a straight line was obtained which describes the rate of penetration into the blastocyst of each of the compounds studied. The intercept of this line on the abscissa corresponds to the reciprocal value of $T_{1/2}$ eq.

RESULTS. *Experiments in Vivo. C^{14}-caffeine.* Six hours after a p.o. dose (3.5 mg/kg; 5.0 μc/kg) of C^{14}-caffeine to three six-day pregnant rabbits, an average of 23.3% (range 13.7–29.4) of the administered label was excreted in their urine. The six-hour urine of two of these rabbits contained an average of 7.9% of the total administered dose as unchanged caffeine, 12.3% as 1-methylxanthine, 2.5% as 1,3-dimethylxanthine plus 1,7-dimethylxanthine and 1.7% as 1,3-dimethyluric acid.

Peak plasma radioactivity was seen four hours after dosing (fig. 1), although it had reached about 98% of this value at two hours; at six hours it was about 96% of the peak value. At this time C^{14} activity was also present in the endometrium, uterine secretion and blastocysts (table 1). The highest level of radioactivity as

compared to maternal plasma was seen in the uterine secretion, where the ratio of C^{14} in uterine secretion/C^{14} in plasma averaged 1.46. Radioactivity in the blastocyst was less, the average ratio of C^{14} in blastocysts/C^{14} in plasma being 0.85.

C^{14}-caffeine in both plasma and blastocysts was the major radioactive compound, being responsible for more than 50% of the radioactivity (table 2). 1,3-Dimethylxanthine, 1,7-dimethylxanthine and 1,3-dimethyluric acid were also found; 1-methylxanthine, however, could be detected only in plasma and not in the blastocyst. The concentration of caffeine, calculated from the specific activity of this compound, was similar in the blastocysts and in the plasma.

H^3-nicotine. An average of 14.5% (range 3.9–36.5) of the administered radioactivity after an i.v. injection of H^3-nicotine (50 μg/kg; 60 μc/kg) to six-day pregnant rabbits was excreted in the urine one hour after its administration. Unchanged H^3-nicotine accounted for 2.45% of the H^3-administered dose in the urine, cotinine 0.6% and demethylcotinine 0.25%.

The plasma H^3 concentration decreased rapidly for 10 minutes to a value approximately 25% less than that at 5 minutes, and then continued to fall slowly during the remainder of the experiment (fig. 2).

Significant amounts of radioactivity were found in plasma, endometrium, uterine secretion and blastocysts one hour after dosing (table 1). The ratio of the tritium activity of endometrium to that of plasma was about 0.9. The uterine secretions of six-day pregnant rabbits appear to concentrate H^3 activity, since radioactivity was about 10 times greater in the uterine secretions than in the plasma. However, H^3 activity in the uterine secretions of nonpregnant rabbits was only slightly higher than that of maternal plasma, with a uterine secretion/plasma concentration ratio of 1.7.

Table 2 shows that the concentration of nicotine in the blastocyst was 4 times greater than that in the plasma, whereas cotinine appeared to be the major metabolite of nicotine in plasma and uterine secretions. Relatively small amounts of demethylcotinine were found in these tissues. At one hour after dosing nicotine accounted for 40.9% of the radioactivity in the uterine secretion; another 21.9% was identified as cotinine and 4.7% as demethylcotinine.

C^{14}-DDT. One hour after the i.v. injection of

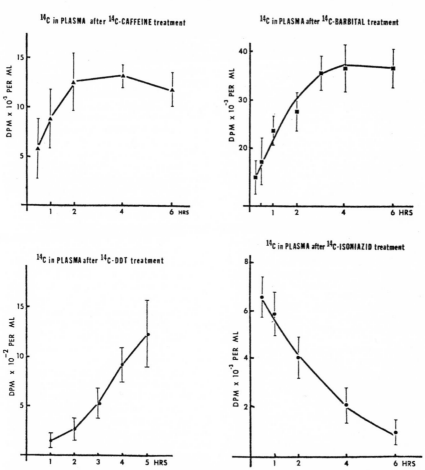

Fig. 1. Radioactivity in the plasma of six-day pregnant rabbits after the p.o. administration of C[14]-labeled caffeine (3.5 mg/kg; 5 μc/kg), barbital (100 ml/kg; 8 μc/kg), DDT (0.7 mg/kg; 5 μc/kg) or isoniazid (13 mg/kg; 5 μc/kg). Points represent the mean value for three animals; bars, standard error of the mean.

C[14]-DDT (400 μg/kg; 3.3 μc/kg) to pregnant rabbits, 3.4% of the administered radioactivity was recovered in the urine. When C[14]-DDT was administered p.o. (700 μg/kg; 5 μc/kg) and the urine collected six hours after treatment, the amount of radioactivity recovered in the urine was 2.5%. Urine collected during a 24-hour period after the p.o. administration of C[14]-DDT (700 μg/kg; 5 μc/kg) to two rabbits contained an average of 8.9% (range 5.1–12.7%) of the total administered radioactivity.

Plasma radioactivity of rabbits treated p.o. with C[14]-DDT rose steadily, and five hours after drug administration had not attained a plateau (fig. 1).

Under the experimental conditions described in table 1, the highest level of tissue radioactivity was seen in the uterine secretions. In the four animals tested, the C[14] in endometrium/C[14] in plasma concentration ratio was highest 6 hours after dosing (2.18) and after 24 hours averaged only 0.67. The ratio of C[14] in blastocyst/C[14] in

TABLE 1

Radioactivity in plasma, endometrium, uterine secretion and preimplantation blastocyst of rabbits treated with C^{14}- or H^3-labeled compounds[a]

| Doe No. | | Treatment | | | | Length of Experiment | Radioactivity in: | | | |
6-day pregnant	Non-pregnant	Radioactive Compound	Route	Dose	Dose		Plasma[b]	Endometrium	Uterine Secretion	Blastocyst
				mg/kg	μc/kg	hr	dpm/ml	dpm/g wet tissue		
5		1-Methyl-C14-caffeine	p.o.	3.5	5	6	16,080	6,030 (0.38)	20,210 (1.26)	11,010 (0.87)
9			p.o.	3.5	5	6	11,724	N.D.	19,450 (1.65)	9,850 (0.84)
25			p.o.	3.5	5	6	14,084	5,810 (0.41)	N.D.	19,927 (0.78)
	80		p.o.	3.5	5	6	17,235	N.D.	25,855 (1.50)	
30		G-H3-nicotine	i.v.	0.05	60	1	178,184	182,382 (1.02)	1,478,971 (8.30)	364,309 (2.04)
31			i.v.	0.05	60	1	203,652	303,483 (1.49)	2,181,449 (10.71)	460,161 (2.26)
36			i.v.	0.05	60	1	141,170	N.D.	1,913,224 (13.55)	222,997 (1.58)
	28		i.v.	0.05	60	1	183,608	215,142 (1.17)	325,674 (1.77)	
	35		i.v.	0.05	60	1	156,110	132,334 (0.85)	221,410 (1.42)	
	37		i.v.	0.05	60	1	197,745	165,141 (0.84)	411,805 (2.08)	
	38		i.v.	0.05	60	1	148,580	156,518 (1.05)	255,967 (1.72)	
	39		i.v.	0.05	60	1	238,040	188,445 (0.79)	437,089 (1.84)	
	40		i.v.	0.05	60	1	192,140	N.D.	326,366 (1.70)	
34		Phenyl-C14-DDT	i.v.	0.4	3.3	1	9,784	8,613 (0.88)	32,553 (3.32)	2,516 (0.25)
29			p.o.	0.7	5	6	1,159	2,527 (2.18)	1,681 (1.45)	137 (0.12)
32			p.o.	0.7	5	24	891	420 (0.47)	3,206 (3.60)	117 (0.13)
	33		p.o.	0.7	5	24	922	896 (0.88)		
1		2-C14-barbital	p.o.	100	10	6	36,027	N.D.	20,640 (0.57)	48,103 (1.34)
16			p.o.	100	8	6	16,544	9,198 (0.56)	23,063 (1.39)	29,209 (1.77)
26			p.o.	100	8	6	17,684	8,711 (0.49)	23,750 (1.34)	27,868 (1.58)
	12		p.o.	30	8	6	16,226	11,033 (0.68)	20,959 (1.29)	
	13		p.o.	30	8	6	16,524	12,698 (0.77)	19,917 (1.21)	
	15		p.o.	100	8	6	14,252	8,836 (0.62)	N.D.	
	23		p.o.	100	8	6	18,718	13,720 (0.73)	20,964 (1.12)	
2		2-C14-thiopental	i.v.	20	10	1	9,285	N.D.	23,924 (3.82)	2,604 (0.27)
24			i.v.	20	10	1	14,864	4,247 (0.28)	11,782 (0.99)	2,254 (0.15)
27			i.v.	20	10	1.5	2,766	2,623 (0.95)	8,633 (3.90)	1,016 (0.37)
21			i.v.	7	7	5	1,938	488 (0.25)	2,796 (1.44)	1,704 (0.88)
22			i.v.	7	7	5	1,066	246 (0.23)	3,659 (3.43)	1,206 (1.13)
	20		i.v.	7	7	5	1,598	622 (0.39)	6,533 (4.09)	
6		Carbonyl-C14-isoniazid	p.o.	13	5	6	1,312	582 (0.44)	2,965 (2.26)	1,704 (1.30)
8			p.o.	13	5	6	1,158	691 (0.06)	2,397 (2.07)	2,008 (1.73)
	4		p.o.	13	5	6	1,290	N.D.	1,642 (1.27)	

[a] For identification of drugs and metabolites associated with the radioactivity values, see table 2.
[b] Figures are mean values of duplicate estimations; those in parentheses are the ratios of radioactivity in tissue/radioactivity in plasma.

plasma was higher (0.25) 1 hour after treatment than at 6 hours (0.12), or at 24 hours (0.13).

C14-barbital. Six hours after the administration of C14-barbital, an average of 13.4% (range 8.2–16.1%) of the C14 was recovered in the urine. Paper chromatographic analysis of the urine with *n*-butanol–acetic acid–water (40:10:50, v/v) as the solvent system indicated that most of the radioactivity (86.3%) was associated with the unchanged compound. However, two additional radioactive zones with R_f values of 0 to 0.15 and 0.35 to 0.45, and accounting for 6.4 and

4.6% of the radioactivity, respectively, were present on the chromatogram. Judging from their chromatographic behavior, it is possible that the two radioactive spots are related to 5-ethylbarbituric acid and 5-ethyl-5(2-hydroxyethyl)-barbituric acid, two known metabolites of barbital in the rat (Goldschmidt and Wehr, 1957).

Plasma radioactivity of rabbits treated p.o. with C14-barbital rose steadily until about three hours after dosing, when the C14 in the plasma reached a maximum level which was maintained

TABLE 2

Radioactive compounds identified in maternal plasma and preimplantation blastocyst of 6-day pregnant rabbits receiving some C¹⁴- or H³-labeled compounds

Compound Administered[a]	Radioactive Compound Identified	Concentration in:	
		Plasma[b]	Blastocyst[c]
		µg/ml	*µg/g*
1-Methyl-C¹⁴-caffeine	Caffeine	2.35 (56.6)[d]	2.49 (71.9)[d]
	1,3-Dimethylxanthine plus 1,7-dimethylxanthine	0.90 (22.7)	0.27 (8.0)
	1-Methylxanthine	0.05 (1.1)	(<0.5)
	1,3-Dimethyluric acid	0.25 (6.2)	0.03 (0.8)
G-H³-nicotine	Nicotine	0.017 (24.2)	0.055 (42.0)
	Cotinine	0.032 (45.5)	0.029 (22.0)
	Demethylcotinine	0.004 (6.4)	0.002 (1.6)
2-C¹⁴-barbital	Barbital	88 (92.4)	117 (89.6)
2-C¹⁴-thiopental	Thiopental	0.26 (38.8)	0.31 (47.3)
Carbonyl-C¹⁴-isoniazid	Isoniazid	0.96 (61.2)	0.97 (48.6)
	Acetylisoniazid	0.20 (12.8)	0.39 (19.5)
	Isonicotinuric acid	0.33 (21.4)	0.35 (17.5)

[a] 1-Methyl-C¹⁴-caffeine (3.5 mg/kg; 5 µc/kg), 2-C¹⁴-barbital (100 mg/kg; 8 µc/kg) and carbonyl-C¹⁴-isoniazid (13 mg/kg; 5 µc/kg) were given by stomach tube, and tissues were examined six hours later. G-H³-nicotine (50 µg/kg; 60 µc/kg) and 2-C¹⁴-thiopental (7 mg/kg; 7 µc/kg) were administered i.v., and tissues were examined one and five hours later, respectively.

[b] Means of the values found for two animals.

[c] Values obtained from pools of 6 to 15 blastocysts.

[d] Figures in parentheses represent the percentage of th e total radioactivity in the tissue.

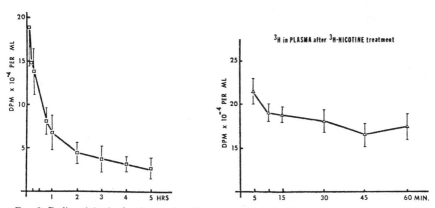

FIG. 2. Radioactivity in the plasma of six-day pregnant rabbits after the i.v. administration of H³-nicotine (50 µg/kg; 60 µc/kg) or C¹⁴-thiopental (7 mg/kg; 7 µc/kg). Points represent the mean value for three animals; bars, standard error of the mean.

throughout the duration of the experiment (fig. 1).

Table 1 shows that the highest level of radioactivity as compared to maternal plasma was present in the free blastocyst, with an average blastocyst C^{14}/plasma C^{14} concentration ratio of 1.56. Radioactivity in the uterine secretion was less, the average C^{14} in uterine secretion/C^{14} in plasma ratio being 1.16. Most of the radioactivity in plasma and in the blastocyst was identified as unchanged C^{14}-barbital (table 2).

C^{14}-thiopental. Five hours after the i.v. injection of C^{14}-thiopental (7 mg/kg; 7 μc/kg), an average of 54.8% (range 32.7–66.6) of the administered radioactivity was excreted in the urine. Unchanged C^{14}-thiopental was identified by paper chromatography, but only 3.1% of the ether-extractable radioactivity was associated with the unchanged compound, whereas the remaining radioactivity was related to other more polar compounds which have not been identified.

Figure 2 shows that plasma radioactivity after treatment with C^{14}-thiopental fell rapidly during the first hour after dosing; after this time, however, the rate of decrease of C^{14} activity slowed so that at six hours plasma radioactivity was still 10% of its value at five minutes after dosing.

Table 1 shows that radioactivity reached the endometrium, uterine secretion and the preimplantation blastocyst after the administration of C^{14}-thiopental to pregnant rabbits. The endometrium contained less radioactivity than the plasma at one and five hours after treatment (ratios of C^{14} in the endometrium/C^{14} in the plasma ranged from 0.23–0.95), whereas the uterine secretion C^{14}/plasma C^{14} concentration ratios ranged from 0.99 to 3.90 (average values of 2.60 and 2.70) one and five hours after dosing, respectively. The blastocyst contained less radioactivity than the plasma; thus, the C^{14} in the blastocyst/C^{14} in the plasma concentration ratios ranged from 0.15 to 0.37 one hour after maternal treatment. By five hours, however, the radioactivity in the blastocyst was approximately equal to that of maternal plasma.

Unchanged thiopental accounted for about 39% of the radioactivity in the plasma and about 47% of that in the blastocyst (table 2).

C^{14}-isoniazid. Six hours after the administration of C^{14}-isoniazid (13 mg/kg; 5 μc/kg) to six-day pregnant rabbits, 61.7% (range 46.1–71.8) of the administered C^{14} was recovered in their urine. The following radioactive compounds, expressed as a percentage of total administered dose, were identified in the urine: unchanged isoniazid, 39.8%; acetylisoniazid, 16.6%; and isonicotinic acid, 6.7%.

C^{14} in the plasma was highest one-half an hour after dosing, and it steadily declined during the remainder of the experiment (fig. 1).

Table 1 shows that radioactivity was present in the endometrium, uterine secretion and preimplantation blastocyst of the treated animals. The ratio of C^{14} activity between endometrium and plasma is about 0.5, whereas the ratios for uterine secretion/plasma and blastocyst/plasma are greater than 1.

Table 2 shows that the preimplantation blastocyst can be penetrated by unchanged isoniazid, as well as by some of its metabolites, acetylisoniazid and isonicotinic acid; the levels of these compounds in the blastocyst are comparable to those in the maternal plasma.

Experiments in Vitro. Figure 3 shows that when blastocysts were incubated in a medium containing C^{14}-dextran of molecular weight 60,000–90,000, radioactivity did not penetrate the blastocysts. Smaller molecular weight C^{14}-dextran (16,000–19,000) penetrated the blastocyst slowly so that after a two-hour incubation C^{14} activity in the blastocyst was not yet in equilibrium with that of the medium. However, the ratio of C^{14} in the blastocyst/C^{14} in the medium reached 50% of the value at equilibrium in a few minutes when blastocysts were incubated in a medium containing either C^{14}-caffeine or C^{14}-barbital. Other compounds with relatively small molecular weights (122–273) entered the blastocyst readily and most obtained $T_{1/2 \ eq}$ in 20 minutes or less (table 3). This table also shows that there is some relationship between both the lipid solubility and degree of ionization of a compound and its rate of uptake by the blastocyst. Thus, the rate of uptake is fast for lipid-soluble compounds, especially those which are predominantly un-ionized at pH 7.2, e.g., thiopental and antipyrine. On the contrary, hexamethonium, which is highly ionized and practically lipid-insoluble, penetrated the blastocyst slowly.

After two hours of incubation *in vitro* with radioactive caffeine, isoniazid, barbital, or salicylic acid only the unchanged compounds could be detected by paper chromatography in the medium and in the blastocyst.

26

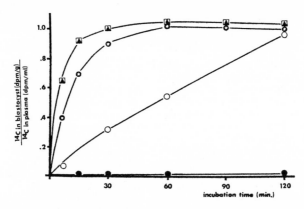

Fig. 3. Rate of appearance of C^{14} activity in preimplantation blastocysts incubated *in vitro* in Ringer-phosphate (pH 7.2) containing C^{14}-caffeine (▲), C^{14}-barbital (◑), C^{14}-dextran of molecular weight 16,000 to 19,000 (○), or C^{14}-dextran of molecular weight 60,000 to 90,000 (●). Points represent the mean values of two experiments in duplicate.

TABLE 3

Rate of uptake by the blastocyst, degree of ionization and lipid solubility of some drugs and other chemicals

Compound[a]	pK_a Values[b]	% Ionized at pH 7.2	Chloroform/Ringer-Phosphate, pH 7.2	$T_{1/2eq}$[c]
Thiopental	7.4 (a) (25)	39	38	1.5
Antipyrine	1.4 (b)	<0.01	23	1.7
Caffeine	0.8 (b) (25)	<0.01	23	2.5
Sulfanilamide	2.4 (c)	<0.01	0.02	6.7
Barbital	7.8 (a) (25)	20	0.30	9.5
Isoniazid	3.8 (d) (20)	0.03	0.04	10.4
Benzoic acid	4.2 (e) (25)	>99	0.39	12.9
Salicylic acid	3.0 (f) (19)	>99	0.01	15.6
Uric acid	3.9 (f) (12)	>99	0.06	20.0
p-Aminobenzoic acid	4.8 (e) (25)	>99	0.01	28.0
Hexamethonium	(Cation)	>99	0.03	66.6

[a] Compounds were incubated at the concentration of 1.5×10^{-3}M.

[b] References for the pK_a values taken from the literature: (a) Bush (1967), (b) Schanker *et al.* (1957), (c) Stecher *et al.* (1968), (d) Salvesen and Glendrange, (1966), (e) Albert and Serjeant (1962), (f) Weast (1962). In parentheses, temperature in degrees centigrade.

[c] Half-equilibrium time—the time (in minutes) which is necessary for a compound, incubated under these conditions, to reach a concentration in the blastocyst equal to half of that at equilibrium.

DISCUSSION. The findings presented in this paper clearly show that compounds administered to the mother are able to penetrate the preimplantation blastocyst after being transferred into the uterine secretion from the maternal circulation. Some compounds reached a concentration in the blastocyst higher than in plasma: H^3-nicotine attained a concentration in blastocyst approximately 4 times higher than in maternal plasma.

For other compounds, *e.g.*, caffeine, barbital, thiopental and isoniazid, the concentration in the blastocyst was approximately equal to that in the maternal plasma. It was found that the radioactivity in the blastocyst, as well as in the plasma, was related not only to the parent compounds, but also to some of their labeled metabolites. In most cases, however, the concentration of each metabolite identified in the blastocyst

27

was less than its concentration in maternal plasma. These metabolites identified in the pre-implantation blastocyst are of maternal origin, since our experiments *in vitro* have shown that blastocysts are not able to form them. In addition, the *in vitro* experiments showed that molecular weight, lipid solubility and the degree of ionization of compounds are important in determining the rate of penetration of compounds into the preimplantation blastocyst.

The amount of compound reaching the blastocyst is dependent upon the transfer of that compound from the maternal circulation into the uterine secretion. All the drugs studied were transferred from the maternal circulation into the uterine secretion in significant amounts. Only nicotine administration resulted in a markedly higher concentration of radioactivity in uterine secretion than in maternal plasma. Most of this radioactivity was identified as unchanged H^3-nicotine, although some of its labeled metabolites were also present. Interestingly, this apparent accumulation of radioactivity was not observed when H^3-nicotine was given to nonpregnant does. This emphasizes the fact that very little is known about the mechanism of excretion of drugs into the uterine secretion, and that there is, to date, no clear picture of how such exocrine glands function. Certainly, factors affecting the plasma level of a drug (dose, route of administration, rate of absorption and excretion, metabolism, tissue localization, protein binding) could influence the transfer of drugs from the maternal circulation into the uterine secretion. Moreover, a direct pharmacologic action of the drug on the uterus, or even an action of the drug on the hormones known to regulate uterine secretory activity, cannot be discounted.

At the present time it is difficult to assess the toxicologic importance of the transfer of drugs and other chemicals into the blastocyst, although there are reports indicating that commonly used drugs may be harmful to the conceptus before implantation (Adams *et al.*, 1961; Lutwak-Mann and Hay, 1962, 1969; Chang, 1964; Schardein *et al.*, 1965).

REFERENCES

ADAMS, C. E., HAY, M. F. AND LUTWAK-MANN, C.: The action of various agents upon the rabbit embryo. J. Embryol. Exp. Morphol. **9**: 468–491, 1961.

ALBERT A. AND SERJEANT, E. P.: Ionization Constants of Acids and Bases, Methuen & Co. Ltd., London, 1962.

ALGERI, E. J. AND WALKER, J. T.: Paper chromatography for identification of the common barbiturates. Amer. J. Clin. Pathol. **22**: 37–40, 1952.

BAKER, J. B. E.: The effects of drugs on the fetus. Pharmacol. Rev. **12**: 37–90, 1960.

BRAMBELL, F. W. R.: The transmission of immunity from mother to young and the catabolism of immunoglobulins. Lancet **2**: 1087–1092, 1966.

BRATTON, A. C. AND MARSHALL, E. K., JR.: A new coupling component for sulfanilamide determination. J. Biol. Chem. **128**: 537–550, 1939.

BUSH, E. T.: General applicability of the channels ratio method of measuring liquid scintillation counting efficiencies. Anal. Chem. **35**: 1024–1029, 1963.

BUSH, M. T.: Sedatives and hypnotics. In Physiological Pharmacology, ed. by W. S. Root and F. G. Hofmann, vol. I, part A, pp. 185–218, Academic Press, New York, 1967.

CHANG, M. C.: Effects of certain antifertility agents on the development of rabbit ova. Fert. Steril. **15**: 97–106, 1964.

COOPER, J. R. AND BRODIE, B. B.: Enzymatic oxidation of pentobarbital and thiopental. J. Pharmacol. Exp. Ther. **120**: 75–83, 1957.

DAVSON, H. AND EGGLETON, M. G., EDITORS: Starling's Human Physiology, 13th ed., p. 39, Lea & Febiger, Philadelphia, 1962.

FABRO, S. AND SIEBER, S. M.: Caffeine and nicotine penetrate the pre-implantation blastocyst. Nature (London) **223**: 410–411, 1969.

FABRO, S., SMITH, R. L. AND WILLIAMS, R. T.: The fate of (^{14}C)-thalidomide in the pregnant rabbit. Biochem. J. **104**: 565–569, 1967.

GARDNER, T. S., WENIS, E. AND LEE, J.: The synthesis of compounds for the chemotherapy of tuberculosis. IV. The amide function. J. Org. Chem. **19**: 753–757, 1954.

GOLDSCHMIDT, S. AND WEHR, R.: Ueber Barbiturate. III. Mitteilung: Der Metabolismus von Veronal. Hoppe-Seyler's Z. Physiol. Chem. **308**: 9–19, 1957.

GOLDSTEIN, A., ARONOW, L. AND KALMAN, S. M.: Principles of Drug Action, pp. 179–194, Harper & Row, Publishers, New York, 1968.

GOLDSTEIN, A. AND WARREN, R.: Passage of caffeine into human gonadal and fetal tissue. Biochem. Pharmacol. **11**: 166–167, 1962.

HUGGETT, A. ST. G. AND HAMMOND, J.: Physiology of the placenta. *In* Marshall's Physiology of Reproduction, 3rd ed., ed. by A. S. Parkes, chap. 16, Longmans, Green & Co. Ltd., London, 1952.

KEBERLE, H., SCHMID, K., FAIGLE, J. W., FRITZ, H. AND LOUSTALOT, P.: Ueber die Penetration von Körperfremden Stoffen in den jungen Wirbeltierkeim. Bull. Schweiz. Akad. Med. Wiss. **22**: 134–152, 1966.

KRISHNAN, R. S. AND DANIEL, J. C.: "Blastokinin": Inducer and regulator of blastocyst development in the rabbit uterus. Science (Washington) **158**: 490–492, 1967.

LUTWAK-MANN, C. AND HAY, M. F.: Effect on the early embryo of agents administered to the mother. Brit. Med. J. **2**: 944–946, 1962.

LUTWAK-MANN, C., HAY, M. F. AND NEW, D. A. T.: Action of various agents on rabbit blastocysts *in vivo* and *in vitro*. J. Reprod. Fert. **18**: 235–257, 1969.

MARKHAM, R. AND SMITH, J. D.: Chromatographic studies of nucleic acids. I. A technique for the identification and estimation of purine and pyrimidine bases, nucleosides and related substances. Biochem. J. **45**: 294–298, 1949.

McKennis, H., Jr., Schwartz, S. L. and Bowman, E. R.: Alternate routes in the metabolic degradation of the pyrrolidine ring of nicotine. J. Biol. Chem. **239**: 3990–3996, 1964.

Moya, F. and Thorndike, V.: Passage of drugs across the placenta. Amer. J. Obstet. Gynecol. **84**: 1778–1798, 1962.

Salvesen, B. and Glendrange, J. H.: The potentiometric determination of the dissociation constants of isoniazid and iproniazid. Medd. Norsk Farm. Selsk. **28**: 272–278, 1966. (Cited by Chem. Abst. **66**: 6498, (68866c) 1967).

Schanker, L. S., Shore, P. A., Brodie, B. B. and Hogben, C. A. M.: Absorption of drugs from the stomach. I. The rat. J. Pharmacol. Exp. Ther. **120**: 528–539, 1957.

Schardein, J. L., Woosley, E. T., Hamilton, L. E. and Kaump, D. H.: Effects of aspirin and phenylbutazone on the rabbit blastocyst. J. Reprod. Fert. **10**: 129–132, 1965.

Sieber, S. M. and Fabro, S.: The penetration of drugs into the rabbit blastocyst before implantation. Pharmacologist **10**: 199, 1968.

Stecher, P. G., Windholz, M. and Leahy, D. S., Editors: The Merck Index, 8th ed., Merck & Co., Inc., Rahway, N.J., 1968.

Villee, C. A.: Placental transfer of drugs. Ann. N.Y. Acad. Sci. **123**: 237–242, 1965.

Wang, C. H. and Willis, D. L.: Radiotracer Methodology in Biological Science, p. 134, Prentice-Hall, Inc., Englewood Cliffs, N.J., 1965.

Weast, C. R.: Handbook of Chemistry and Physics, 45th ed., The Chemical Rubber Publishing Co., Cleveland, 1962.

Yale, H. L., Lossee, K., Martins, J., Holsing, M., Perry, F. M. and Bernstein, J.: Chemotherapy of experimental tuberculosis. VIII. The synthesis of acid hydrazides, their derivatives and related compounds. J. Amer. Cancer Soc. **75**: 1933–1942, 1953.

Zamboni, V. and Defranceschi, A.: Identification of isonicotinoylhydrazones of pyruvic acid and alpha-ketoglutaric acid in rat urine after treatment with isonicotinic acid hydrazide (isoniazid). Biochim. Biophys. Acta **14**: 430–431, 1954.

29

DDT Administered to Neonatal Rats Induces Persistent Estrus Syndrome

W. L. Heinrichs, R. J. Gellert
J. L. Bakke, N. L. Lawrence

The hypothalamus of developing rats apparently establishes its neural mechanisms for regulating gonadotropin secretion in either a female (cyclic) or male (tonic) pattern during a critical period early in postnatal life (*1*). The direction of this neural development is toward the male pattern if the perinatal gonad secretes male hormone, and toward the female one in the hormone's absence. Little information is available about the critical period of neural development in those species whose offspring are not so immature at birth. In them that period probably occurs earlier, during intrauterine life. Exposure of newborn female rats to various exogenous estrogens or androgens induces permanent sterility with polycystic ovaries, anovulation, persistent vaginal estrus, and absence of female mating behavior (*1*). Clomiphene, 1-[*p*(*β*-diethylaminoethoxy)phenyl]-1,2-diphenyl-2-chloroethylene, a synthetic agent with weak estrogenic properties, similarly produces permanent sterility (*2*). Uterotropic effects of the insecticide DDT [1,1,1-trichloro-2-(*o*-chlorophenyl)-2-(*p*-chlorophenyl)-ethane] have recently been described (*3*). The *o,p′*-isomer of DDT, which comprises approximately 20 percent of technical grade DDT, is several times more uterotropic than *p,p′*-DDT. In view of the

Table 1. Effects of *o,p′*-DDT administered to neonatal rats. All figures are means and standard errors. Numbers of rats are shown in parentheses.

Treat-ment	Age (days)		Weight (mg/100 g body weight)			Lordosis quotient	
	Vaginal opening	First estrus	Ovarian	Uterine		Test 1	Test 2
				Oil	Estradiol		
Control	33.9 ± 0.6 (13)	34.9 ± 0.9 (13)	31.1 ± 2.0 (8)	34.6 ± 1.8 (4)	141.4 ± 5.1 (4)	69 ± 13 (8)	65 ± 12 (8)
DDT	30.9 ± 0.9* (12)	31.5 ± 1.0† (12)	25.2 ± 1.8‡ (12)	39.5 ± 3.1 (6)	110.4 ± 5.9* (6)	51 ± 14 (12)	50 ± 12 (12)

* Different from controls ($P < .01$). † $P < .02$. ‡ $P < .05$.

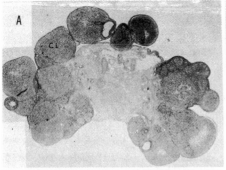

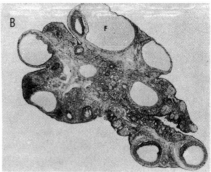

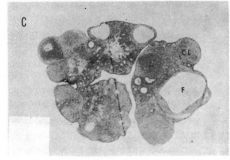

Fig. 1 (A). Section of ovary from the control group. Note many large corpora lutea (*CL*). (B and C) Two examples of sections taken from animals treated with DDT. Note the cystic follicles (*F*) and absence of corpora lutea in (B). Section (C) also has cystic follicles but some small corpora lutea.

ubiquity and persistence of chlorinated hydrocarbons and their ability to reach the fetus (*4*), we investigated the effects of *o,p'*-DDT injected into neonatal female rats because of the sensitivity of the neural development to estrogen and the distinctive character of the resultant syndrome.

Sprague-Dawley rats were obtained from timed matings, and birth days were designated as day zero. Female pups were pooled and randomly assigned to a control group or one treated with DDT. The DDT (1 mg) was administered subcutaneously on the 2nd, 3rd, and 4th days of life; control animals received only the propylene glycol and ethanol vehicle. The animals were given free access to Purina Lab Chow and tap water and had a schedule of artificial diurnal lighting with 14 hours of light. All rats were weaned at 21 days of age and caged in pairs. After vaginal opening was noted, daily smears were obtained by vaginal lavage. At 120 days of age, ovariectomy was performed under ether anesthesia, the ovaries were weighed, and one from each animal was fixed for histological examination. The remaining ovaries were pooled and frozen for analysis of DDT. Mating behavior was tested 19 days later after the animals had been primed with a standard alternating schedule of estradiol benzoate and progesterone. Tests were repeated in the following weeks. Responses were scored as lordosis quotients, the ratio of the number of lordoses per ten mounts by a male (*5*). Seventeen days after the second mating test we injected one-half of each group with 10 μg of free estradiol in sesame oil daily for 1 week to evaluate uterine sensitivity to estrogen. The remaining animals received only the oil vehicle. Twenty-four hours after the last injection, the animals were decapitated and the uteri were weighed. The whole brain and samples of suprarenal adipose tissue were frozen and submitted along with the ovaries to analysis of DDT residues by means of gas-liquid chromatography (6).

Significant advances in vaginal opening and first estrus were the first indications that DDT treatment during the neonatal period altered the reproductive system (Table 1). Estrous cycles were normal in the treated group (12 animals) until approximately 60 days of age, when the first signs of persistent vaginal estrus appeared (no less than four consecutive days with a cornified vaginal smear). By 100 days of age, all of the treated rats showed the first signs of persistent estrus, and by 120 days, when ovariectomy was performed, persistent cornification was established in all rats.

Ovaries of the treated group contained large cystic follicles and a marked reduction in fresh corpora lutea (Fig. 1). These histologic findings were reflected in the differences in ovarian weight, which were of borderline statistical significance (Table 1).

When the persistent estrus syndrome is marked by a delay in the onset of anovulation it is termed the delayed anovulatory syndrome. The incidence of anovulation in animals treated with testosterone as neonates increases with time (7). Since the high incidence of cornified vaginal smears was occasionally interrupted by 1 or 2 days of diestrus and the differences in ovarian weight of the groups had only a borderline statistical significance, it is likely that our animals were in the early stages of the development of this syndrome. Had the observation period in this experiment been extended, a longer absence of ovulation and greater regression of existing corpora lutea would likely have yielded a more dramatic diminution in ovarian weight.

Indirect evidence suggests that the anovulation is due to damage to hypothalamic mechanisms regulating cyclic secretion of luteinizing hormone (LH). Anovulatory animals produced by neonatal treatment with testosterone ovulate after administration of LH, electrical stimulation of the preoptic area of the hypothalamus (8), or administration of LH releasing factor (9). Furthermore, the syndrome is prevented by simultaneous administration of depressive agents of the central nervous system such as chlorpromazine and barbiturates (10). The advancement of puberty may be due to an altered hypothalamic sensitivity regulating gonadotropin secretion; however the mechanism is unknown. The presence of a persistently cornified vaginal smear without the usual cyclic diestrous appearance of the leukocytes is probably due to the tonic secretion of estrogen as well as to the absence of progesterone which normally attends ovulation.

Female rats treated during the neonatal period with sex steroids often fail to show lordosis and female mating behavior (5). In two such mating tests, the treated animals scored lower than controls, but the differences were not statistically significant (Table 1). Neonatal neural mechanisms are more sensitive to the induction of the persistent estrus syndrome than are those controlling behavior (11). The high dose of estrogen increased uterine weight in both the control and treated groups (Table 1); however the increment in the latter group was significantly reduced. We presume that the injection schedule for estrogen and progesterone used in tests for mating behavior increased the uterine weights of control and treated rats; however the increments could not be measured.

Table 2. DDT residues in tissues of rats treated with o,p'-DDT as neonates. Unless data are given, concentrations of o,p'-DDT, p,p'-DDE, and p,p'-DDD in the tissues were below the level of sensitivity (0.01 ppm) of the technique. Each value represents a single determination from a pool size shown in parentheses.

Residue	Tissue	Amount (ppm)	
		Control	DDT
p,p'-DDT	ovary	0.03	0.03
		(4)	(7)
p,p'-DDT	brain	0.01	0.01
		(4)	(6)
p,p'-DDT	adipose	0.30	0.34
p,p'-DDE	adipose	0.18	0.18
		(4)	(6)

Nevertheless, the similarity in initial weights of both groups injected with oil indicates that the degree of uterine regression was similar in both groups after cessation of that treatment. Harris (12) also noted reduced uterine responses to estrogen in animals treated with sex steroids during the neonatal period. The reduced sensitivity may be due to decreased estradiol receptor (13), but the mechanism whereby the receptor is permanently reduced is unknown. Preliminary data obtained by sucrose density gradient centrifugation suggest that the concentration of estradiol receptor in the uterine cytosol from the treated animals is less than controls (14).

It is surprising that the amounts of DDT residues in ovary, brain, and adipose tissue (Table 2) were not affected by treatment of neonates with DDT. Purina Lab Chow contained 0.01 part of p,p'-DDT per million which probably accounts for the residues found in all the animals. The similar amounts in the treated and control groups indicate that the injected DDT had been cleared before the au-topsy. Therefore, the alteration of uterine response in vivo and in vitro cannot be accounted for by residual DDT, although the amounts in uterine tissue were not measured. The o,p'-DDT isomer was not detected in any of the tissues, probably because it is converted to the p,p'-isomer in living tissues (15).

These data show that DDT, in addition to sex steroids, can induce the constant estrus syndrome with permanent sterility. The similarity between this syndrome in rats and the polycystic ovary syndrome found in women has been described (16). The oligo-ovulation and relative sterility in the human syndrome could be related to the presence of DDT in the fetal environment (4), but the relationship remains to be established.

References and Notes

1. G. W. Harris, Endocrinology 75, 627 (1964).
2. W. W. Leavitt and D. M. Meismer, Nature 218, 181 (1968); R. J. Gellert, J. L. Bakke, N. Lawrence, Fertil. Steril. 22, 244 (1971).
3. J. Bitman, H. C. Cecil, S. J. Harris, G. F. Fries, Science 162, 371 (1968); R. M. Welch, W. Levin, A. H. Conney, Toxicol. Appl. Pharmacol. 14, 358 (1969).
4. J. A. O'Leary, J. E. Davis, W. F. Edmundson, G. A. Reich, Am. J. Obstet. Gynecol. 107, 65 (1970); Z. W. Polishuk, M. Wassermann, D. Wassermann, Y. Groner, S. Lazarovici, L. Tomatis, Arch. Environ. Health 20, 215 (1970).
5. H. H. Feder and R. E. Whalen, Science 147, 306 (1965).
6. We thank Dr. J. Allard and A. L. Robbins of the Washington State Division of Health for analyzing the DDT residues in our samples. The technique, a modification of that reported by H. F. Enos, F. J. Biros, D. T. Gardner, and J. P. Wood [Abstr. Amer. Chem. Soc. Meet. 154th (1967)], has a sensitivity of 0.01 ppm for the compounds named in Table 2.
7. R. A. Gorski, Endocrinology 82, 1001 (1968).
8. C. A. Barraclough and R. A. Gorski, ibid. 68, 68 (1961).
9. Y. Arai, J. Fac. Sci. Univ. Tokyo 10, 243 (1963).
10. ——— and R. A. Gorski, Endocrinology 82, 1005 (1968).

11. C. A. Barraclough and R. A. Gorski, *J. Endocrinol.* **25**, 175 (1962).
12. G. W. Harris and S. Levine, *J. Physiol. (London)* **181**, 379 (1965).
13. B. Flerkó and B. Mess, *Acta Physiol. Acad. Sci. Hung.* **33**, 111 (1968); J. L. McGuire and R. D. Lisk, *Nature* **221**, 1068 (1969).
14. M. Sarff, R. J. Gellert, W. L. Heinrichs, unpublished results.
15. M. C. French and D. J. Jefferies, *Science* **165**, 914 (1969).
16. K. B. Singh, *Obstet. Gynecol. Surv.* **24**, 2 (1969).
17. We thank Dr. Mary Sarff for measuring the concentration of the uterine receptor; Dr. Harvey Feder, Department of Biology, Rutgers University for aid in the sexual receptivity studies; and Susan Robinson, Jane Bennett, and G. Comito for technical assistance. Supported by American Cancer Society, Inc., grant T-543 and PHS grants HD-03825 and AM-05638-09. Portions of this study were presented at the 1971 Annual Meeting of the Western Society for Clinical Research, Carmel, Calif.; they have been published in abstract form [W. L. Heinrichs, R. J. Gellert, J. L. Bakke, N. L. Lawrence, *Clin. Res.* **19**, 171 (1971)].

Metabolic Alterations in the Squirrel Monkey Induced by DDT Administration and Ascorbic Acid Deficiency

R. W. CHADWICK, M. F. CRANMER, AND A. J. PEOPLES

The growing threat to human welfare presented by increasing world population, inadequate nutrition, and environmental pollution could be further complicated by various interactions between nutritional status and toxic stress.

Primates and guinea pigs, unlike other mammals, cannot synthesize ascorbic acid. Ascorbic acid has been implicated in the control of oxido-reduction states of living cells, but the detailed role of vitamin C has not yet been fully outlined (Staudinger et al., 1961). However, there have been numerous reports indicating that ascorbic acid deficiency decreases the in vitro activity of various drug-metabolizing enzymes in guinea-pigs (Axelrod et al., 1954; Conney et al., 1961; Degkwitz and Staudinger, 1965; Degkwitz et al., 1968; Leber et al., 1969; Kato et al., 1969; Wagstaff and Street, 1971). Together with glucuronyl transferase, the drug-metabolizing enzymes represent an important means of detoxifying both foreign and endogenous lipid soluble compounds.

In mammals, agents which induce drug-metabolizing enzymes generally stimulate the excretion of L-ascorbic acid via the glucuronic acid system. This has been taken as evidence that there may be a biochemical link between the induction of drug-metabolizing enzyme systems and the stimulation of the glucuronic acid system (Aarts,

1968). There have been no reports regarding the effect of ascorbic acid deficiency on the enzymes of the glucuronic acid system. Also, no evidence exists to indicate that primates, as well as guinea pigs, show impaired enzyme activity in response to ascorbic acid deficiency.

This study was conducted to investigate possible effects of ascorbic acid deficiency and repeated exposure to DDT on various detoxification routes in the squirrel monkey, *Saimiri seireus*.

METHODS

Twenty-four female squirrel monkeys *Saimiri seireus*, were assigned to 1 of 4 treatment groups in a 2×2 factorial experiment arranged in randomized blocks involving DDT and ascorbic acid. It has been reported that the chlorinated hydrocarbon insecticides induce optimum enzyme activity in guinea pigs when these animals receive 200–250 mg/kg of ascorbic acid (Wagstaff and Street, 1971). In order to obtain a similar level for the squirrel monkey, all animals were maintained for 2 wks on an ascorbic acid deficient diet fortified with 3000 ppm ascorbic acid.[2] During the third and fourth wk half of the animals continued on the same diet while the other half received the diet without the ascorbic acid supplement. Six monkeys from the 12 fed diets containing ascorbic acid and 6 from the 12 on the deficient diet received, in addition, daily po administration of 5 mg of *p,p'*-DDT in peanut oil. The remaining 12 monkeys were given peanut oil po. After 14 days all animals received po 5 mg of DDT and 5 mg of γHCH (containing 0.15 μCi of ^{36}Cl-DDT and 4.6 μCi of ^{14}C-γHCH).[3,4] All animals were housed in individual metabolism cages to enable the quantitative collection of urine and feces.

Throughout the experiment, 24-hr urine samples were collected and analyzed daily for both total D-glucuronic acid and D-glucaric acid. Total glucuronic acid was determined colorimetrically by a slight modification of the Kanobrocki method (E. Kanobrocki, personal communication). Urine samples were diluted to 100 ml and 0.1-ml aliquots were added to 12-ml glass stoppered test tubes. Following the addition of 1 ml of technical grade hydrochloric acid[5] and 1 ml of 0.2% naphthoresorcinol,[6] the tubes were placed in a 100°C heating block for 30 min. The tubes were then immersed in ice water for 10 min after which 5 ml of anhydrous ethyl ether was added. The tubes were shaken for 30 sec, and the colored ether layer was decanted and analyzed on a Coleman Jr. II at 540 nm using a water blank to zero the instrument. The 0.2% naphthoresorcinol solution was prepared daily, 1 day in advance of the analysis. After dissolving 0.2 g of naphthoresorcinol in 100 ml of water, oxygen was bubbled through the solution for 1 hr, and it was placed in the refrigerator overnight. Before use the next day, nitrogen was bubbled through the solution for 1 hr. A standard curve using glucuronolactone[7] was assayed daily along with the urine. The D-glucaric acid was determined by an enzymic assay (Marsh, 1963a).

Fifteen days after treatment began, the animals were sacrificed. Urine, feces, fat, liver

[2] The ascorbic acid-deficient diet and ascorbic acid were obtained from Nutritional Biochemicals Corporation, Cleveland, Ohio.
[3] The ^{36}Cl-DDT was obtained from Mallinckrodt Nuclear, Orlando, Florida.
[4] The ^{14}C-γHCH was obtained from Nuclear Chicago, Des Plaines, Illinois.
[5] Technical grade hydrochloric acid was obtained from Mallinckrodt Chemical Works, St. Louis, Missouri.
[6] Naphthorisorcinol was obtained from Nutritional Biochemicals Corporation, Cleveland, Ohio.
[7] Glucuronolactone was obtained from Nutritional Biochemicals Corporation, Cleveland, Ohio.

and brain samples were taken for analysis of radioactivity. Liver samples were assayed for cytochrome P-450 (Alvares *et al.*, 1968), microsomal protein (Gornall *et al.*, 1949) and the in vitro activity of various hepatic enzymes. The enzyme systems studied included those responsible for the O-demethylation of p-nitroanisole, (Kinoshita *et al.*, 1966), the oxidative hydrolysis of O-ethyl-O-(4-nitrophenyl) phenylphosphonothioate (EPN), (Kinoshita *et al.*, 1966), the reduction of the azo group of sodium p-dimethyl-aminoazobenzenesulfonate (Methyl Orange), (Fouts *et al.*, 1957), and the glucuronidation of p-nitrophenol (Chadwick *et al.*, 1971). The epoxidation of aldrin to dieldrin was also investigated (Gillett *et al.*, 1966).

The weight of the monkeys, liver wet weight, total liver protein (Gornall *et al.*, 1949) and glycogen content (Murphy and Porter, 1966) were determined. Lipid content was determined by extracting tissue samples with hexane, and after vaporizing the solvent, weighing the residue.

Dried samples of feces and liver were analyzed for radioactivity by the oxygen flask combustion method (Davidson and Oliverio, 1967). Radioactivity of the hexane extracts of brain and adipose tissue was also determined. Urine samples were analyzed for ^{14}C and ^{36}Cl radioactivity. Radioactivity of all samples was determined with a Nuclear-Chicago Mark I liquid scintillation counter.

Analysis of variance for a 2×2 factorial experiment arranged in randomized blocks was used as an aid in the interpretation of the data produced in this experiment (Cochran and Cox, 1957). Because of the number of tests involved it was decided that 8 was the optimum number of animals which could be sacrificed per day. Thus the monkeys were housed in 3 racks of 8 cages each. Each rack contained 2 monkeys on each of the 4 treatments. Treatment was begun on each rack of animals on consecutive days. After the 14-day treatment period, the racks of monkeys were sacrificed in the same sequence and the randomized block design was used to keep the experimental error within each group as small as practicable.

Analysis of the trend in the urinary excretion ratio of D-glucuronic/D-glucaric acid was utilized (von Neumann, 1941; Gart, 1942). Designed comparisons (Snedecor, 1956) and correlation coefficients (Snedecor, 1956) were also employed to aid in the interpretation of these data. Comparisons are considered significantly different at $p \leqslant 0.05$ unless otherwise stated.

RESULTS

The effect of the 2-wk treatment on various detoxification routes in the squirrel monkey was evaluated by a variety of measured responses: (1) glucuronic acid and glucaric acid excretion, (2) excretion and storage of radioactivity, (3) liver weight, liver cytochrome P-450 content and liver microsomal protein content and (4) the in vitro activity of a variety of hepatic drug-metabolizing enzymes.

Glucuronic Acid and Glucaric Acid Excretion

The sequence of biochemical processes leading from D-glucose or D-galactose to D-glucuronic acid and then to D-glucaric, L-ascorbic and D-xylulose is referred to as the glucuronic acid system (Aarts, 1968). Glucuronidation is an important detoxification mechanism in mammals, and there is some evidence that a biochemical link between the induction of drug-metabolizing enzyme systems and the stimulation of the glucuronic

37

acid system may exist (Aarts, 1968). Thus the effects of DDT stress and ascorbic acid deficiency on the urinary excretion of total (including free and conjugated) D-glucuronic acid and D-glucaric acid were recorded daily throughout this experiment.

All the treatment groups showed a gradually increasing D-glucaric acid content in the urine from a low on day 3 to about twice the amount on day 15, (Fig. 1). During the final wk of the study, the average glucaric acid excretion of the monkeys fed the ascorbic acid deficient diet was significantly higher than that of the animals fed the diet fortified with 3000 ppm ascorbic acid. There was no significant difference between the 2 groups treated with DDT (Table 1).

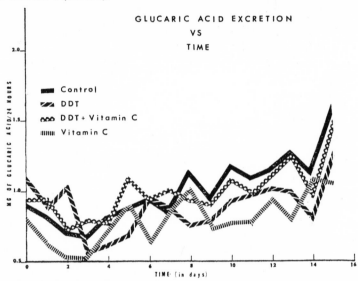

FIG. 1. Effect of DDT and ascorbic acid on the excretion of D-glucaric acid (mg/24-hr urine sample). Each point represents the mean excretion value of 6 animals. The treatment groups designated control and DDT received an ascorbic acid-deficient diet.

Animals receiving ascorbic acid and DDT + ascorbic acid excreted significantly more D-glucuronic acid on days 5, 7 and 14 than other animals in the study (Fig. 2). The DDT + ascorbic acid treated monkeys excreted significantly more total glucuronic acid during the last week of treatment than the animals receiving DDT only (Table 1).

Since marked individual differences in drug metabolism within the same species have been observed (Burns, 1968), the ratio of the 24-hr D-glucuronic/D-glucaric acid excretions for each monkey was computed and the average treatment values were plotted (Fig. 3). Monkeys receiving ascorbic acid excreted a significantly higher ratio of D-glucuronic/D-glucaric acid than the animals on the deficient diet by the second day of treatment. This higher excretion ratio was maintained throughout the experiment with the exception of day 8. The general decline in the D-glucuronic/D-glucaric acid ratio from day 8 to the end of the experiment is of interest in view of the fact that on day 7 a new lot of the ascorbic acid deficient diet was opened and administered to all animals. The animals were maintained on this feed for the remainder of the experiment.

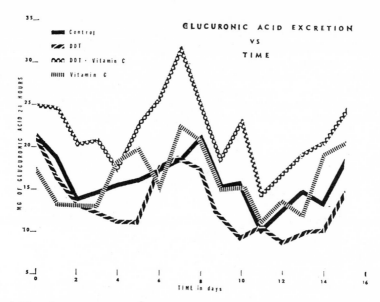

FIG. 2. Effect of DDT and ascorbic acid on the excretion of D-glucuronic acid (mg/24-hr urine sample). Each point represents the mean excretion value of 6 animals. The treatment groups designated control and DDT received an ascorbic acid-deficient diet.

TABLE 1

AVERAGE EXCRETION OF D-GLUCARIC AND D-GLUCURONIC ACID
DURING THE FINAL WEEK OF THE EXPERIMENT

Treatment		Average urinary excretion/24 hr	
Ascorbic acid (ppm)	DDT (mg)	Glucaric acid[a] (mg, mean ± SE)	Glucuronic acid[a] (mg, mean ± SE)
0.0	0.0	1.17 ± 0.10	14.6 ± 1.4
0.0	5.0	0.96 ± 0.05	11.1 ± 0.8
3000	0.0	0.88 ± 0.06	15.4 ± 1.2
3000	5.0	1.10 ± 0.08	19.7 ± 1.4

Analysis of variance			
	df	Mean square	Mean square
Controls vs. DDT treated	1	0.000	7.714
Ascorbic acid-fed controls vs. ascorbic acid-deficient controls	1	1.82[b]	16.30
DDT + ascorbic acid vs. DDT	1	0.412	1569[b]
Error	162	0.232	58.26
Blocks	2	1.50[b]	472.5[b]
Total	167	0.256	71.71

[a] The mean is the average of 7 determinations on each of 6 monkeys.
[b] $p < 0.05$.

The D-glucuronic/D-glucaric acid excretion pattern was tested for a significant trend before and after the administration of the new lot of feed (von Neumann, 1941; Hart, 1942). The decline in the excretion ratio during the second half of the study constituted a significant trend at $p < 0.05$ while no trend was detected for the first 8 days. That the trend was similar for all groups was supported by a 93 % correlation between the excretion ratio of the ascorbic acid treated animals and that of the ascorbic acid deficient group. Similarly the excretion ratio of the monkeys receiving DDT + ascorbic acid and that of the DDT-treated animals has a correlation coefficient of 81 %.

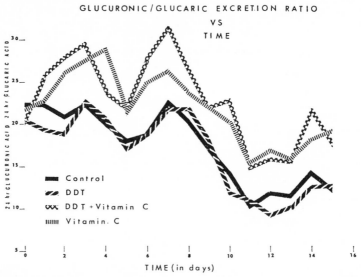

FIG. 3. Effect of DDT and ascorbic acid on the D-glucuronic/D-glucaric acid excretion ratio (mg D-glucuronic acid/mg D-glucaric acid). Each point represents the mean value of 6 animals. The treatment groups designated control and DDT received an ascorbic acid-deficient diet.

Excretion and Storage of Radioactivity

The excretion and storage of ^{14}C-γHCH derived material is summarized in Table 2. The DDT-treated monkeys excreted significantly more ^{14}C-radioactivity in the feces and significantly less in the urine than the animals not treated with the insecticide. The DDT-treated monkeys also stored significantly less γHCH in the liver and fat than the other animals based on analysis of radioactivity in these tissues. Though somewhat depressed in DDT-treated monkeys, the level of γHCH in brain lipid was not significantly lower than that of the other animals.

The average excreted ^{14}C-radioactivity for the DDT-treated monkeys amounted to 38 % of the administered dose while that of the other animals was 31 % of the administered dose.

The μCi of ^{36}Cl-DDT administered on day 14 was not large enough to permit detection of ^{36}Cl-radioactivity in either urine or brain tissue. There were no significant treatment differences in the level of ^{36}Cl-radioactivity found in feces, fat or liver. All animals excreted about 60 % of the administered dose of ^{36}Cl-radioactivity in the feces.

TABLE 2

STORAGE AND EXCRETION OF ^{14}C-RADIOACTIVITY

Treatment		Excreted ^{14}C[a]		Stored γHCH[b]		
Ascorbic acid (ppm)	DDT (mg)	Urine	Feces	Liver[c]	Fat[d]	Brain[d]
0.0	0.0	2.72 ± 0.56	29.9 ± 3.7	3.72 ± 1.07	26.5 ± 3.1	6.2 ± 1.0
0.0	5.0	1.60 ± 0.38	38.9 ± 5.4	2.26 ± 0.49	18.9 ± 3.9	5.0 ± 1.0
3000	0.0	2.50 ± 0.79	28.3 ± 4.8	3.65 ± 0.57	23.9 ± 2.4	6.1 ± 0.7
3000	5.0	1.71 ± 0.26	36.0 ± 2.0	2.43 ± 0.74	16.2 ± 3.0	4.7 ± 0.8

Analysis of variance						
Source	df	Mean square	Mean square	Mean square	Mean square	Mean square
DDT effect	1	5.41[e]	418[e]	10.8[e]	349[f]	10.7
Ascorbic acid effect	1	0.0187	29.9	0.0123	40.2	0.240
DDT × ascorbic acid	1	0.159	2.41	0.0820	0.0193	0.0417
Blocks	2	3.11	1.16	4.94	26.8	4.89
Error	18	1.56	116	3.22	63.6	4.71
Total	23	1.73	111	3.42	69.0	4.59

[a] Average DPM × 10^{-5} ± SE of the mean.
[b] Average ppm γHCH based on ^{14}C-radioactivity present.
[c] Average ppm γHCH in tissue sample.
[d] Average ppm γHCH in tissue lipid.
[e] $P < 0.10$.　　[f] $P < 0.05$.

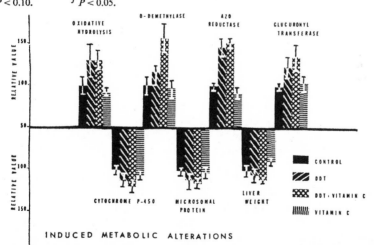

FIG. 4. Effect of DDT and ascorbic acid on enzyme activity, cytochrome P-450 content, hepatic microsomal protein content, and liver weight. For each parameter the mean value of the monkeys receiving neither DDT nor ascorbic acid is represented as 100% ± SE. The mean values of the other treatment groups relative to that of these ascorbic acid-deficient controls are plotted as percent ± SE. (Treatment mean/control mean × 100.) Each bar represents the mean value of 6 monkeys. The treatment groups designated control and DDT received an ascorbic acid-deficient diet.

Enzyme Activity and Metabolic Alterations

The data from this part of the study are summarized in Fig. 4. Relative values of all parameters are plotted with 100% representing the animals receiving neither ascorbic acid nor DDT. Only statistically significant data ($p < 0.05$) are present in Fig. 4.

Monkeys treated with DDT, when compared to the other animals in the study, had significantly larger livers and significantly higher levels of both cytochrome P-450 and microsomal protein, but no significant differences in body weight or hepatic glycogen were observed.

The activities of the enzyme systems responsible for the oxidative hydrolysis of EPN, the reduction of the azo group of methyl orange and the glucuronidation of *p*-nitrophenol were all significantly elevated in the animals treated with DDT. Induction of the *O*-demethylation of *p*-nitroanisole by DDT was significantly enhanced by the presence of ascorbic acid in the diet while no significant treatment differences in the epoxidation of aldrin to dieldrin were noted.

DISCUSSION

A new oxidative metabolic pathway which leads to the formation of D-glucaric acid has been demonstrated in mammals (Marsh, 1963a, b, c). The physiological significance of this new pathway lies in the production of the endogenous β-glucuronidase inhibitor, D-glucaro-$(1 \rightarrow 4)$lactone, which plays an important role in controlling glucuronide hydrolysis (Levvy and Conchie, 1966). Since many foreign and endogenous substances are detoxified and excreted as glucuronides, the action of D-glucaro-$(1 \rightarrow 4)$ lactone in arresting the liberation of active toxic moieties by β-glucuronidase may serve as one of the most important defense mechanisms.

The total D-glucuronic acid content of the urine can be influenced by a number of different factors including glucuronyl transferase induction, hydrolysis of glucuronides by β-glucuronidase and stimulation of the formation of D-glucuronic acid itself. For example, it is apparent when interpreting the results of this study that dietary ascorbic acid is essential to obtain a maximum D-glucuronic/D-glucaric acid excretion ratio (Fig. 3). However, the mechanism involved in this requirement appears to be quite complex.

Thus, while the D-glucaric acid excretion by the ascorbic acid-deficient DDT-treated animals is relatively low, in vitro glucuronyl transferase activity is significantly higher than that of the controls. Excretion of small amounts of D-glucaric acid is indicative of a low antagonism to the hydrolysis of biliary glucuronides. However, the increased glucuronyl transferase would tend to counteract any increase in hydrolysis so that neither would significantly alter the excretion of total D-glucuronic acid. The average D-glucuronic acid content of the urine from the ascorbic acid-deficient DDT-treated monkeys is significantly lower than that of the animals receiving DDT and ascorbic acid during the final week of the study (Table 1). It appears that ascorbic acid deficiency may impair the stimulation of the D-glucuronic acid system by DDT and thereby contribute to a low D-glucuronic/D-glucaric acid excretion ratio.

On the other hand, the ascorbic acid-treated monkeys excrete significantly less D-glucaric acid during the last week than the controls on the deficient diet (Table 1) without a corresponding increase in glucuronyl transferase (Fig. 4). Accordingly the resultant increased hydrolysis may contribute significantly to a high D-glucuronic/D-glucaric acid excretion ratio from the ascorbic acid treated monkeys.

Repeated DDT treatment is known to stimulate the D-glucuronic acid system in rats leading to enhanced production of both ascorbic acid (Street *et al.*, 1966) and D-glucuronic acid (Chadwick *et al.*, 1971). In rats, there is evidence that while glucuronide synthesis is insensitive to the various metabolites of the glucuronic acid pathway both lysosomal and microsomal β-glucuronidase are strongly inhibited by D-glucaric acid (Hanninen, 1968). Unlike other mammals, primates and guinea pigs cannot synthesize ascorbic acid and a marked decrease of in vitro drug metabolizing enzyme activity has been well documented for scorbutic guinea pigs (Axelrod *et al.*, 1954; Conney *et al.*, 1961; Degkwitz and Staudinger, 1965; Degkwitz *et al.*, 1968; Leber *et al.*, 1969; Kato *et al.*, 1969; Wagstaff and Street, 1971). However, nothing has been published regarding the effect of ascorbic acid deficiency on the enzymes of the glucuronic acid system in these animals.

In addition to ascorbic acid deficiency, an undetermined nutritional factor from the administration of a second lot of the ascorbic acid deficient diet appeared to effect an inhibition of the D-glucuronic/D-glucaric acid excretion ratio. In the rat, both the quality and the quantity of dietary protein influence the excretion of ascorbic acid, a metabolite of the glucuronic acid system in these animals (Chandrasekhara *et al.*, 1968). The need for further investigations of the effects of both toxic stress and nutritional status on the enzymes of the glucuronic acid system is indicated.

Although induction of a number of drug metabolizing enzymes in guinea pigs is markedly impaired within 14 days of ascorbic acid deficiency (Wagstaff and Street, 1971), only *O*-demethylase appears to be similarly affected in the squirrel monkey (Fig. 4). The fact that the guinea pig metabolizes ascorbic acid at a very rapid rate may account for this discrepancy (Burns, 1968). The unusual sensitivity of the *O*-demethylase enzyme system in both primates and guinea pigs is noteworthy. Squirrel monkeys chronically treated with 0.5 mg/kg DDT po had *O*-demethylase levels significantly higher than the controls. However, in the same animals no induction of the enzyme responsible for oxidative hydrolysis of EPN could be detected (Cranmer *et al.*, 1971). A similar phenomenon has been observed in guinea pigs. In addition, while the in vitro level of *O*-demethylase was significantly increased prior to the onset of scorbutic change in guinea pigs, the activity of the enzymes detoxifying EPN were not significantly altered by several known enzyme inducers (J. C. Street, personal communication). In rats, on the other hand, the enzymes responsible for the oxidative hydrolysis of EPN are more sensitive to induction by DDT than the *O*-demethylase system. After 3 wk on diets containing 25 ppm DDT, the in vitro activity of *O*-demethylase increased 42% while the oxidative hydrolysis of EPN rose 84% (Kinoshita *et al.*, 1966). It may be that *O*-demethylase is a more important and responsive component of the detoxification capabilities of primates and guinea pigs than of the rat.

While DDT seems to exert some influence over the storage and excretion of γHCH, the effect is much smaller than that observed in rats (Chadwick *et al.*, 1971). Ascorbic acid deficiency did not seem to affect the metabolism of γHCH. The squirrel monkey may not absorb γHCH from the gut as readily as the rat since preliminary GLC analysis indicates that considerable quantities of unaltered γHCH are excreted in the feces of the monkey.

Neither DDT pretreatment nor ascorbic acid deficiency appear to influence the storage and excretion of DDT.

43

Results of this study suggest that interactions between nutritional and toxic stress may significantly affect some routes of detoxification in the squirrel monkey. Furthermore, the D-glucuronic/D-glucaric acid excretion ratio may serve as a sensitive early warning signal of possible nutritional deficiency and enzyme impairment in primates.

REFERENCES

AARTS, E. M. (1968). Drug-induced stimulation of the glucuronic acid system, pp. 1, 5. Doctoral Thesis, Drukkerij Gebr. Janssen N.V., Nijmegen.

ALVARES, A. P., SCHILLING, G., LEVIN, W., and KUNTZMAN, R. (1968). Alterations of the microsomal hemoprotein by 3-methyl-cholanthrene: Effects of ethionine and actinomycin D. *J. Pharmacol. Exp. Ther.* **163**, 417–424.

AXELROD, J., UDENFRIEND, S., and BRODIE, B. B. (1954). Ascorbic acid in aromatic hydroxylation (III). Effect of ascorbic acid on hydroxylation of acetanilide, aniline and antipyrine. *J. Pharmacol. Exp. Ther.* **111**, 176–181.

BURNS, J. J. (1968). Variation of drug metabolism in animals and the prediction of drug action in man. *Ann. N. Y. Acad. Sci.* **151**, 959–967.

CHADWICK, R. W., CRANMER, M. F., and PEOPLES, A. J. (1971). Comparative stimulation of γHCH metabolism by pretreatment of rats with γHCH, DDT, and DDT + γHCH. *Toxicol. Appl. Pharmacol.* **18**, 685–695.

CHANDRASEKHARA, N., RAO, M. V. L., and SRINIVASAN, M. (1968). Influence of dietary protein on the enzyme system synthesizing ascorbic acid in the rat. *Indian J. Biochem.* **5**, 22–24.

COCHRAN, W. G., and COX, G. M. (1957). *Experimental Designs*, pp. 148–182. Wiley, New York.

CONNEY, A. H., BRAY, G. A., EVANCE, C., and BURNS, J. J. (1961). Metabolic interactions between L-ascorbic acid and drugs. *Ann. N. Y. Acad. Sci.* **92**, 115–127.

CRANMER, M., PEOPLES, A., and CHADWICK, R. (1971). Some effects on squirrel monkeys on repeated administration of *p,p'*-DDT. *Toxicol. Appl. Pharmacol.* in press.

DAVIDSON, J. D., and OLIVERIO, V. T. (1967). Tritium and carbon-14 by oxygen flask combustion, Atomlight No. 60, May 1967, New England Nuclear Corp., Boston, Mass. See also: *Anal. Biochem.* **4**, 188–189 (1962).

DEGKWITZ, E., and STAUDINGER, H. (1965). Untersuchungen zur Hydroxylierung von Acetanilid mit Lebermikrosomen normaler und skorbutischer Meerschweinchen. *Hoppe-Seyler's Z. Physiol. Chem.* **342**, 63–72.

DEGKWITZ, E., LUFT, D., PFEIFER, U., and STAUDINGER, H. (1968). Untersuchungen über mikrosomale Enzymaktivitaten (cumarinhydroxylierung), NADPH-Oxidation, Glucose-6-Phosphatase und Esterase) und Cytochromgehalte (P-450 und b_5) bei normalen, skorbutischen und hungernden Meerschweinchen. *Hoppe-Seyler's Z. Physiol. Chem.* **349**, 465–471.

FOUTS, J. R., KAMM, J. J., and BRODIE, B. B. (1957). Enzymic reduction of prontosil and other azo dyes. *J. Pharmacol. Exp. Ther.* **120**, 291–300.

GILLETT, J. W., CHAN, T. M., and TERRIERE, L. C. (1966). Interaction between DDT analogs and microsomal exposidase systems. *J. Agr. Food Chem.* **14**, 540–545.

GORNALL, A. G., BARDAWILL, C. J., and DAVID, M. M. (1949). Determination of serum proteins by means of the biuret reaction. *J. Biol. Chem.* **177**, 751–766.

HANNINEN, O. (1968). On the metabolic regulation in the glucuronic acid pathway in the rat tissues. *Ann. Acad. Sci. Fenn. Chem.*, Ser AII, No. 142, 7–96.

HART, B. I. (1942). Significance levels for the ratio of the mean square successive difference to the variance. *Ann. Math. Statist.* **13**, 445–447.

KATO, R., TAKANAKA, A., and OSHIMA, T. (1969). Effect of vitamin C deficiency on the metabolism of drugs and NADPH-linked electron transport system in liver microsomes. *Jap. J. Pharmacol.* **19**, 25–33.

KINOSHITA, F. K., FRAWLEY, J. P., and DuBOIS, K. P. (1966). Quantitative measurement of induction of hepatic microsomal enzymes by various dietary levels of DDT and toxaphene in rats. *Toxicol. Appl. Pharmacol.* **9**, 505–513.

44

LEBER, V. H., DEGKWITZ, E., and STAUDINGER, H. (1969). Untersuchungen zum Einfluss der Ascorbinsäure auf die Aktivität und die Biosynthese mischfunktioneller Oxygenasen sowie den Gehalt an Hamoproteiden in der Mikrosomenfraktion der Meerschweinchenleber. *Hoppe-Seyler's Z. Physiol. Chem.* **350,** 439–445.

LEVVY, G. A., and CONCHIE, J. (1966). In: *Glucuronic Acid Free and Combined* (G. J. Dutton, ed.), pp. 333–349. Academic Press, New York.

MARSH, C. A. (1963a). Metabolism of D-glucuronolactone in mammalian systems. I. Identification of D-glucaric acid as a normal constituent of urine. *Biochem. J.* **86,** 77–86.

MARSH, C. A. (1963b). Metabolism of D-glucuronolactone in mammalian systems. II. Conversion of D-glucaric acid by tissue preparations. *Biochem. J.* **87,** 82–90.

MARSH, C. A. (1963c). Metabolism of D-glucuronolactone in mammalian systems. III. Further studies of D-glucuronolactone dehydrogenase of rat liver. *Biochem. J.* **89,** 108–114.

MURPHY, S. D., and PORTER, S. (1966). Effects of toxic chemicals on some adaptive liver enzymes, liver-glycogen and blood glucose in fasted rats. *Biochem. Pharmacol.* **15,** 1665–1676.

SNEDECOR, G. W. (1956). *Statistical Methods Applied to Experiments in Agriculture and Biology,* pp. 254, 160. Univ. Press, Ames, Iowa.

STREET, J. C., CHADWICK, R. W., WANG, M., and PHILLIPS, R. L. (1966). Insecticide interactions affecting residue storage in animal tissues. *J. Agr. Food Chem.* **14,** 545–549.

STAUDINGER, H., KRISCH, K., and LEONHAUSER, S. (1961). Role of ascorbic acid in microsomal electron transport and the possible relation to hydroxylation reactions. *Ann. N. Y. Acad. Sci.* **92,** 195–207.

VON NEUMANN, J. (1941). Distribution of the ratio of the mean square successive difference to the variance. *Ann. Math. Statist.* **12,** 367–395.

WAGSTAFF, D. J., and STREET, J. C. (1971). Ascorbic acid deficiency and induction of hepatic microsomal hydroxylative enzymes by organochlorine pesticides. *Toxicol. Appl. Pharmacol.* **19,** 10–19.

45

Effects of DDT and of Drug–DDT Interactions on Electroshock Seizures in the Rat

DOROTHY E. WOOLLEY

The pharmacologic interaction between DDT and drugs or other commonly used chemicals is of theoretical and practical interest (Brodie *et al.*, 1965). Medically, it is important to be able to predict possible synergisms and antagonisms between DDT and chemicals with which man comes in contact, because measurable quantities of DDT have been demonstrated in tissues of people all over the world (Hayes, 1959; Hayes *et al.*, 1963). Also, interactions between DDT and drugs or toxins having fairly well-understood modes of action may provide further understanding of the mechanism of action of DDT.

The present study of DDT–drug interactions in the rat used the durations of phases of the maximal electroshock seizure (MES) and of postseizure depression as the end points. The MES test for investigating CNS effects of drugs and toxicants is simple and readily quantifiable and has the potential of providing more information on mechanisms of action than determination of the LD50, for example. Most convulsant and anticonvulsant drugs modify the MES pattern in a predictable way; e.g., convulsant drugs increase, and anticonvulsant drugs decrease, the duration of tonic extension. Because convulsions are one of the symptoms of DDT poisoning in most species (Hayes, 1959), DDT also would be expected to modify the phases of the MES.

The effects of acute administration of DDT on the MES phases were first compared with the effects of three prototype drugs, two of which were convulsant, i.e., pentylenetetrazol and strychnine, and one of which was anticonvulsant, i.e., pentobarbital. Then the effects on the MES of each of the three drugs in combination with DDT were determined. The three drugs were selected for comparison with DDT because their actions at the level of the whole nervous system and at the neuronal level have been thoroughly studied, and because each is representative of a different type of drug action. Pentylenetetrazol is an "axon unstabilizer" like DDT and veratrum (Eyzaguirre and Lilienthal, 1949; Shanes, 1951), strychnine blocks postsynaptic inhibition (Eccles, 1965), and pentobarbital is a general CNS depressant.

METHODS

Appropriate doses and testing intervals were determined in preliminary studies. About 250 female Sprague-Dawley rats were used in the studies reported here. Rats weighed 180–210 g for studies on the effects of pentobarbital, strychnine and pentylenetetrazol, 200–220 g for the DDT study, and 220–250 g for the DDT–drug interaction studies. The procedure and apparatus for producing the MES were similar to those described by Woodbury and Davenport (1952). Briefly, silver disk electrodes with a small amount of electrode paste were placed on the scalp in front of the ears and behind the eyes, approximately over the medial portion of each temporal cortex, and a sine wave (60 cps) stimulus of 200 mA was delivered for 0.2 sec by a constant current stimulator. Four electronic timers were activated simultaneously with the shock to the animal and were turned off manually at the end of the phases of the seizure. A stopwatch was used to time postseizure depression. After the convulsion the animal was placed on its side in a small box (8 × 10 × 5 in. high) attached to the platform of a tilting apparatus and rocked from side to side at the rate of 28 times per minute. Postseizure depression was considered ended when the animal had recovered the righting reflex (RRR) sufficiently to be able to remain standing in the box. About 20% of the rats tested did not show tonic extension during the seizure following the 200 mA electroshock, and these animals were not used. Rats were used for testing only if they demonstrated tonic extension during the MES on each of three different days.

Ten to 12 rats were used in each control and treatment group. Rats were fasted for 12 hours before drugs or insecticides were administered. DDT [2,2-bis(p-chlorophenyl)-1,1,1-trichloroethane; 99% p,p'-isomer; m.p., 110.5°C] was dissolved in cottonseed oil (Wesson) and delivered by stomach tube. Control rats were intubated with the oil vehicle. Pentylenetetrazol (Metrazol, Knoll Pharmaceutical Co.), pentobarbital (Diabutal, Diamond Laboratories), and strychnine sulfate (Merck) were diluted with or dissolved in physiological saline and were injected subcutaneously in a volume of about 0.2 ml 1 hour before eliciting the MES. Control rats were injected with the saline vehicle.

For the DDT–drug combination studies, DDT or the oil vehicle was given 12 hours and the drug or saline vehicle 1 hour before the MES was determined.

Pretreatment mean values for 0 dose (control) and experimental groups were analyzed (*t* test, nonpaired data) to assure that significant differences between groups did not exist before treatment. Pre- and posttreatment values for each group were compared (*t* test, paired data) to evaluate the effects of repeated shocking itself and the reproducibility of the MES, as well as the effects of the drugs or pesticides.

RESULTS

Among the motor responses elicited in control rats by the electroshock were: first, tonic flexion of the hindlimbs for 3–5 sec; second, full extension lasting about 6–7 sec

PENTOBARBITAL

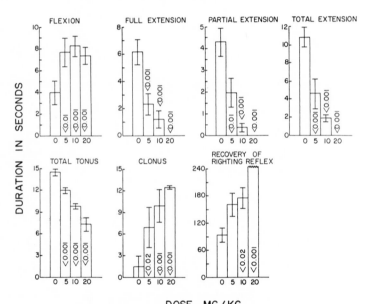

FIG. 1. Effects of 3 doses of pentobarbital in the rat on the duration of phases of the maximal electroshock seizure (MES) and of the postseizure depression measured as the time to recover the righting reflex. The MES was measured 1 hour after the subcutaneous injection of the drug or saline. The vertical bracketed lines represent the standard errors of the means. Pretreatment values were determined 48 hours before posttreatment values, which are shown in the figure. *P* values, written on or above each barogram, are based on the *t* test for paired data and refer to the significance of the differences between the means for pretreatment and posttreatment values for the 0 dose and 3 drug-treated groups.

when the hindlimbs were tonically extended; and partial extension for about 4 sec when the hindlimbs were tonically contracted in a position intermediate between full extension and complete flexion. Clonus did not occur in most control rats. Usually, 1–2 min were required to recover the righting reflex after the convulsion.

DDT and Drugs Alone

The MES of the 0 dose groups did not change significantly during the 48 hours between obtaining pre- and posttreatment values in any of the 3 drug studies, and only the posttreatment values are shown (Figs. 1–3). Thus, the MES of the controls was stable and reproducible under these conditions. No deaths occurred in the 3 series of drug experiments.

The doses of pentobarbital investigated for their effects on the MES were well below the dose for surgical anesthesia (35 mg/kg) in the rat. Each of the three doses (5, 10, and

STRYCHNINE

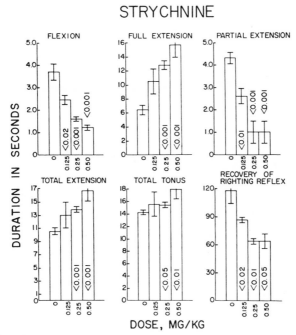

FIG. 2. Effects of 3 doses of strychnine on the duration of phases of the MES in the rat. Conditions are the same as in Fig. 1.

20 mg/kg) significantly increased the duration of flexion to about the same extent (Fig. 1). The two lower doses markedly decreased duration of the extensor phases, and the highest dose abolished extension in each animal. Total duration of the tonic phases was progressively decreased with increasing dose. Treatment with pentobarbital increased the occurrence and duration of clonus and increased postseizure depression Each rat treated with the highest dose of pentobarbital required more than 4 min to recover the righting reflex.

The doses of strychnine (0.125, 0.25, and 0.50 mg/kg) used did not cause spontaneous convulsions. With increasing doses durations of flexion, partial extension and recovery of the righting reflex were decreased and durations of full extension, total extension, and total tonus were increased (Fig. 2).

The three doses of pentylenetetrazol (30, 40, and 50 mg/kg) did not significantly alter the duration of tonic flexion or recovery of the righting reflex after the seizure, whereas durations of full extension, partial extension, total extension and total tonus increased with increasing dose (Fig. 3).

The effects of 50, 100, and 200 mg/kg DDT were determined 6, 12, 24, 48, and 120 hours after administration (Table 1). Although the LD50 for DDT was about 250 mg/kg in unshocked rats, only 2 of the 10 rats treated with 200 mg/kg and 6 of the 10 given 100 mg/kg were still alive for the MES and RRR at 24 hours. All of the 10 controls and the

PENTYLENETETRAZOL

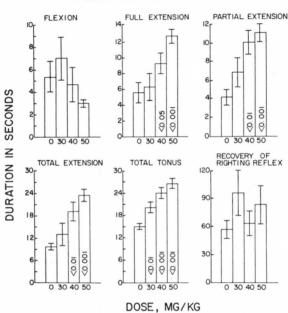

FIG. 3. Effects of 3 doses of pentylenetetrazol on the durations of phases of the MES in the rat. Conditions are the same as in Fig. 1.

10 rats given 50 mg/kg DDT survived. The mortality is attributed to the stress effects of the combination of the MES and DDT. The principal gross behavioral changes observed were tremors and hyperexcitability at 6 and 12 hours in all groups.

The principal effects of DDT on the seizure pattern occurred at 6 and 12 hours and were an increase in the duration of tonic flexion and a decrease in the durations of full extension, total extension, total tonus, and RRR. The effects of 100 mg/kg were long-lasting and showed evidence of being biphasic. After 24 hours, duration of flexion was still significantly increased and durations of the extensor phases were still decreased in this group. After 120 hours the opposite effects were observed: duration of tonic flexion was decreased and full extension was increased. In general, control values were stable during the testing period, but showed some significant changes from pretreatment values,

50

TABLE 1
EFFECTS OF DDT ON DURATION OF PHASES OF THE MAXIMAL ELECTROSHOCK SEIZURE IN THE RAT[a]

Dose (mg/kg)	Hours after DDT					
	−24	+6	+12	+24	+48	+120
Flexion (sec)						
0	3.23	4.20	4.78	4.31	3.30	5.03
	±0.34	±1.35	±1.32	±1.05	±0.55	±1.74
50	3.38	6.28*	6.94**	5.18	5.03	3.81
	±0.37	±1.46	±1.14	±1.27	±1.39	±1.27
100	3.40	6.03*	7.53*	8.66**	2.90	2.27**
	±0.13	±1.45	±1.68	±1.71	±0.21	±0.17
200	3.34	5.85*	7.81***	—	—	—
	±0.30	±1.19	±1.01			
Full extension (sec)						
0	7.43	6.65	5.69	5.46*	6.36	7.32
	±0.36	±0.85	±1.04	±1.00	±0.81	±1.25
50	6.58	3.11**	2.66**	5.36	5.11	8.09
	±0.56	±0.90	±1.10	±1.08	±1.23	±1.01
100	7.89	3.86***	2.40***	1.86***	6.77**	10.20
	±0.49	±1.11	±1.17	±1.24	±0.39	±0.61
200	7.05	3.24**	0.63***	—	—	—
	±0.60	±0.98	±0.63			
Partial extension (sec)						
0	4.80	3.95	3.59	3.55	3.86	3.31
	±0.45	±0.51	±0.65	±0.65	±0.49	±0.62
50	4.81	3.25	2.37*	3.47	2.98**	3.75
	±0.21	±0.92	±0.99	±0.64	±0.71	±0.51
100	4.28	2.90	1.53*	1.48*	3.15	2.93
	±0.58	±0.82	±0.76	±0.97	±0.37	±0.52
200	4.99	3.18	1.26**	—	—	—
	±0.40	±0.88	±0.83			
Total extension (sec)						
0	12.23	10.90	9.27	9.19*	10.22	10.63
	±0.56	±1.26	±1.59	±1.59	±1.20	±1.79
50	11.69	6.36**	5.03**	8.78*	8.10**	11.84
	±0.63	±1.74	±2.06	±1.62	±1.80	1.40
100	12.20	6.77**	4.28**	3.35***	9.92***	13.13
	±0.59	±1.86	±1.70	±2.12	±0.53	±0.54
200	11.94	6.32**	2.40***	—	—	—
	±0.46	±1.75	±1.60			
Total tonus (sec)						
0	15.55	14.70	14.05	13.49*	13.52*	15.67*
	±0.53	±0.30	±0.83	±0.74	±0.72	±0.45
50	15.07	12.56***	11.00**	13.96	13.02*	15.65
	±0.37	±0.42	±1.14	±0.63	±0.69	±0.31
100	15.68	12.80***	11.69	11.96	12.82	15.40
	±0.61	±0.77	±0.73	±1.06	±0.68	±0.58
200	15.29	12.17	9.26	—	—	—
	±0.49	±0.74	±0.89			

51

TABLE 1—*continued*

Dose	Hours after DDT					
(mg/kg)	−24	+6	+12	+24	+48	+120
	Recovery of righting reflex (min)					
0	2.01	2.09	1.78	2.24	2.42	2.63**
	±0.33	±0.32	±0.32	±0.39	±0.41	±0.35
50	1.85	1.22*	1.33	2.12	2.15	2.32
	±0.21	±0.12	±0.24	±0.41	±0.26	±0.18
100	1.78	1.44*	1.18*	1.95	2.23	2.65
	±0.13	±0.28	±0.28	±0.51	±0.48	±0.76
200	1.70	1.16**	1.59	—	—	—
	±0.18	±0.06	±0.21			

[a] Values are means ± SE. *$P < 0.05$, **$P < 0.02$, ***$P < 0.001$, based on *t* test for paired data, i.e., pretreatment versus posttreatment values for each rat.

especially 24 hours after administration of the vehicle. These minor changes may be attributed to effects of the repeated shocking.

Further information on the effects of DDT is presented in Fig. 4, in which the duration of full extension is plotted against the duration of flexion before and 12 hours after

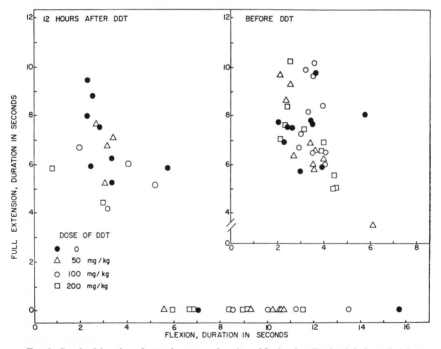

FIG. 4. Graph of duration of extension versus duration of flexion for all animals before administration of DDT (inset, right) and 12 hours after DDT.

administration of DDT. Pretreatment values showed a negative correlation ($r = -0.51$; $P < 0.01$), so that increasing durations of flexion were associated with shorter durations of full extension. The correlation between *total* extension and flexion during the pretreatment period was similar ($r = -0.46$; $P < 0.01$). After administration of DDT, animals showed increased duration of flexion, and many did not exhibit extension. The percent of animals showing extension during the posttreatment period for the various groups was as follows: 80% for the 0 mg/kg DDT group; 40% for the 50 mg/kg group; 50% for the 100 mg/kg group; and 22% for the 200 mg/kg group. Extension did not occur in animals in which duration of flexion was greater than about 5 sec. A similar increase in duration of flexion and lack of occurrence of extension was produced by pentobarbital (Fig. 1).

DDT–Drug Interactions

Investigation of the effects of 75 mg/kg DDT and 40 mg/kg pentylenetetrazol, alone and in combination, on the MES showed that the combination was highly toxic. Six of the 12 rats treated with both died within an hour after pentylenetetrazol administration, i.e., less than 12 hours after DDT administration. Deaths occurred before the MES could be determined and so could not be attributed to the additional stress of the MES.

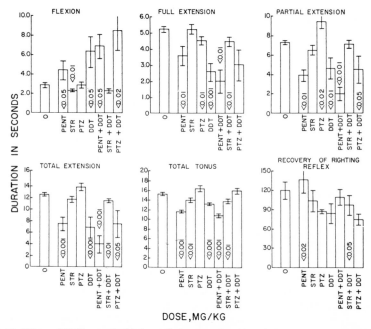

FIG. 5. Effects of DDT, pentobarbital, strychnine, pentylenetetrazol, alone and in combination, on the durations of phases of the MES and the postseizure depression in the rat. DDT was given 12 hours and the drugs 1 hour before the MES was determined. Conditions are the same as in Fig. 1. Abbreviations and doses are as follows: PENT = pentobarbital, 5 mg/kg; STR = strychnine, 0.125 mg/kg; PTZ = pentylenetetrazol, 30 mg/kg; DDT = DDT, 50 mg/kg.

No deaths occurred in the groups receiving DDT or pentylenetetrazol only. The rats remaining in the DDT-pentylenetetrazol group were tested and showed an MES pattern basically similar to that shown by the group having received DDT only, i.e., increased duration of tonic flexion and decreased durations of full and total extension. Interestingly, administration of pentylenetetrazol to DDT-treated animals increased the duration of tonic flexion slightly more than it was increased in rats with DDT only. Pentylenetetrazol alone hastened recovery of the righting reflex in this experiment, but not in the preceding study (Fig. 3).

The effects of low doses of DDT, pentobarbital, strychnine and pentylenetetrazol alone and of the DDT–drug combinations are shown in Fig. 5. Pentobarbital alone increased the duration of flexion and postseizure depression, and decreased the duration of the extensor phases and total tonus. The pattern produced by DDT was similar to that of pentobarbital except for recovery of the righting reflex. The combination of DDT and pentobarbital generally resulted in greater effects on the seizure phases, especially on partial extension, than did either agent alone. DDT pretreatment prevented the increase in duration of postseizure depression normally produced by pentobarbital.

The only significant effect of strychnine alone was to decrease the duration of tonic flexion. However, strychnine given to DDT-treated rats either prevented or lessened some effects produced by DDT alone, i.e., the increase in duration of the flexor phase and the decrease in duration of the extensor phases.

Pentylenetetrazol alone decreased the duration of full extension and increased the duration of partial extension. When given to DDT-treated rats, pentylenetetrazol produced a greater increase in the duration of tonic flexion than occurred in rats having received DDT only. The increase in duration of flexion was significant at the 1 % level for rats having received both DDT and pentylenetetrazol, whereas it was significant only at the 5 % level for rats having received only DDT. Pentylenetetrazol reduced the effect of DDT in shortening the duration of the extensor phases and of total tonus.

DISCUSSION

In rats the clinical symptoms and changes in brain electrical activity during DDT poisoning clearly show that DDT is excitatory to the CNS. Hyperexcitability, as evidenced by an exaggerated startle response to sudden noises, tremors, and convulsions are symptoms of DDT poisoning in most species, including the rat (Hayes, 1959). In awake, unrestrained rats with chronically implanted electrodes, many of the changes in brain electrical activity produced by DDT administration were similar to changes occurring during behavioral arousal (Woolley and Barron, 1968). Hence, the effects of DDT on the MES phases were unexpected because they were similar to the effects of pentobarbital and were largely opposite to those of the two excitatory, convulsant drugs. Like pentobarbital, DDT increased the duration of flexion and decreased the duration of the extensor phases and of total tonus. The combination of DDT and pentobarbital produced even greater effects on the MES phases, whereas the combinations of DDT and pentylenetetrazol or strychnine resulted in lesser effects, than did DDT alone. The problem then is to explain why DDT and pentobarbital have similar effects on the pattern of maximal electroshock convulsions.

54

The effects of pentobarbital on the MES pattern are believed to be due primarily to actions on the spinal cord rather than on the brain (Esplin, 1959; Esplin and Freston, 1960). Pentobarbital prolongs presynaptic inhibition in the spinal cord (Eccles *et al.*, 1963; Eccles, 1965), and this effect is probably sufficient to account for the MES changes produced. Normally, the extensor motoneurons are under a greater degree of background inhibition than are the flexor motoneurons. Hence, during a maximal seizure the first response of the hindlimbs is tonic flexion, and only the flexor motoneurons are active. During flexion, facilitation of extensor motoneurons occurs and extensors begin to fire. Extension results despite continued activity of the flexors, because extensor muscles are more powerful (Esplin and Laffan, 1957; Esplin and Freston, 1960). Increased inhibition on extensor motoneurons would prolong flexion by increasing the time required to cause firing of extensors, and this appears to be the mechanism of action of pentobarbital.

By contrast, strychnine removes background inhibition in the spinal cord by blocking postsynaptic inhibition (Eccles, 1965). Low doses of strychnine shorten flexion and high doses abolish it during electroshock seizures (Fig. 2; Esplin and Zablocka, 1965). Pentylenetetrazol does not block either presynaptic or postsynaptic inhibition (Eccles, 1965); in the spinal cord its excitatory effects are partially counteracted because it stimulates both inhibitory and excitatory systems whereas its effects on the brain are only excitatory (Lewin and Esplin, 1961; Esplin and Zablocka, 1965).

The present findings are consistent with the hypothesis that DDT increases duration of tonic flexion by increasing inhibition at the spinal level. Although this hypothesis requires direct test before it can be accepted, the following observations support this view: (1) the similarity in the effects of DDT and pentobarbital and the known actions of pentobarbital in increasing spinal inhibition; (2) the effectiveness of strychnine in preventing the increased duration of flexion caused by DDT and the known actions of strychnine in decreasing spinal inhibition; and (3) the slight additive effects of DDT and pentylenetetrazol on duration of flexion (Fig. 5) plus the evidence that pentylenetetrazol increases activity in both spinal inhibitory and excitatory systems, which suggests that the additive effects of the two agents on spinal inhibition are not completely overcome by excitatory effects. A similarity in effects between DDT and pentylenetetrazol was expected because both have "veratrinic" actions on the nervous system (Eyzaguirre and Lilienthal, 1949). Their opposite effects on duration of extension become reconciled when it is recognized that in the intact animal the MES effects of pentylenetetrazol are due to actions on the brain rather than on the spinal cord (Lewin and Esplin, 1961), whereas the reverse appears to be true for the MES effects of DDT.

The hypothesis that the DDT-induced increase in duration of flexion results from a primary action on the spinal cord receives support from previous observations that the hindlimbs of cats with thoracic spinal transection demonstrated augmentation of flexor tone and lowering of the threshold for the flexion reflex (Bromiley and Bard, 1949). A direct action of DDT on the spinal cord of the rat has also been observed by Shankland (1964). Also, comparison of the fresh tissue concentrations of DDT in the rat shows that concentrations are as high or higher in the spinal cord than in the brain during 24 hours after acute administration of DDT (Woolley and Runnells, 1967).

The negative correlation between durations of flexion and extension (Fig. 4) appears to be valid for durations of flexion between 2 and 5 sec and is consistent with the hypothesis that duration of flexion is directly related to the degree of inhibition of extensor

55

motoneurons. When duration of flexion is greater than about 5 sec, as after DDT or pentobarbital treatment, extension is not likely to occur, suggesting that the period during which activation of the extensor motoneurons may take place lasts for only about 5 sec after the electroshock.

The well-known effect of DDT in increasing activity of the liver microsomal drug-metabolizing enzymes (Conney, 1967) would not affect the present results on DDT–drug interactions, because the increased activity does not occur within 24 hours after DDT. It is less well known that an acute dose of DDT may cause an immediate *decrease* in activity of enzymes that metabolize barbiturates and, consequently, may increase tissue levels of pentobarbital and prolong hexobarbital and pentobarbital sleeping time in rats and mice (Hart and Fouts, 1963; Sung and Chow, 1966). This effect may have contributed to the additive effects of DDT and pentobarbital in the present study.

Although the effects of DDT on the MES were similar to the effects of pentobarbital, the significant shortening of RRR 6 hours after DDT was in contrast to the marked prolongation of RRR caused by pentobarbital. The faster RRR after DDT administration may be related to the transient period of hyperglycemia caused by DDT (Hayes, 1959). High blood sugar levels have been reported to shorten postseizure depression (Timiras *et al.*, 1956), presumably by helping replenish the metabolic substrates depleted during the seizure. A shortening of neuronal recovery time by DDT, as suggested by Eyzaguirre and Lilienthal (1949), although questioned by Shankland (1964), also would be expected to shorten postseizure depression.

The inconsistent effect of pentylenetetrazol on RRR observed here has also been noted by others (Castillo, 1964) and may be due to multiple effects which counteract each other. Pentylenetetrazol acts to increase the intensity of the MES and this alone would increase the severity of the postseizure depression and prolong RRR. On the other hand, pentylenetetrazol also stimulates respiration (Woolley *et al.*, 1965) and shortens neuronal recovery time (Eyzaguirre and Lilienthal, 1949), which would hasten RRR.

The unexpected effects of DDT and of the DDT–drug combinations on the maximal seizure pattern in rats emphasize that the actions of DDT on the CNS still require further study. The present results provide a firmer basis for predicting interactions between certain types of drugs and DDT. The results also show that the MES pattern provides a sensitive test for detecting CNS effects of DDT.

ACKNOWLEDGMENTS

This study was supported by NIH grant ES-00163. The technical assistance of Dr. Margaret Nesselrod, Miss Cheryl Scott, and Mr. Allan Runnells is gratefully acknowledged.

REFERENCES

BRODIE, B. B., COSMIDES, G. J., and RALL, D. P. (1965). Toxicology and the biomedical sciences. *Science* **148**, 1547–1554.

BROMILEY, R. G., and BARD, P. (1949). Tremor and changes in reflex status produced by DDT in decerebrate, decerebrate-decerebellate, and spinal animals. *Johns Hopkins Hosp. Bull.* **84**, 414–429.

CASTILLO, L. S. (1964). Electroconvulsive responses to convulsant and anticonvulsant drugs during acclimatization at high altitude (12,470 ft) in rats. Master's Thesis. Biology Library, Univ. of California, Berkeley.

CONNEY, A. H. (1967). Pharmacological implications of microsomal enzyme induction. *Pharmacol. Rev.* **19**, 317–366.

ECCLES, J. C. (1965). Pharmacology of central inhibitory synapses. *Brit. Med. Bull.* **21**, 19–25.

ECCLES, J. C., SCHMIDT, R., and WILLIS, W. D. (1963). Pharmacological studies on presynaptic inhibition. *J. Physiol. (London)* **168**, 500–530.

ESPLIN, D. W. (1959). Spinal cord convulsions. *Arch. Neurol.* **1**, 485–490.

ESPLIN, D. W., and FRESTON, J. W. (1960). Physiological and pharmacological analysis of spinal cord convulsions. *J. Pharmacol. Exptl. Therap.* **130**, 68–80.

ESPLIN, D. W., and LAFFAN, R. J. (1957). Determinants of flexor and extensor components of maximal seizures in cats. *Arch. Intern. Pharmacodyn. Therap.* **113**, 189–202.

ESPLIN, D. W., and ZABLOCKA, B. (1965). Central nervous system stimulants. I. In: *The Pharmacological Basis of Therapeutics* (L. S. Goodman and A. Gilman, eds.), 3rd ed., pp. 345–353. Macmillan, New York.

EYZAGUIRRE, C., and LILIENTHAL, J. L., JR. (1949). Veratrinic effects of pentamethylenetetrazol (Metrazol) and 2,2-bis (*p*-chlorophenyl) 1,1,1-tri-chloroethane (DDT) on mammalian neuromuscular function. *Proc. Soc. Exptl. Biol. Med.* **70**, 272–275.

HART, L. G., and FOUTS, J. R. (1963). Effects of acute and chronic DDT administration on hepatic microsomal drug metabolism in the rat. *Proc. Soc. Exptl. Biol. Med.* **114**, 388–392.

HAYES, W. J., JR. (1959). Pharmacology and toxicology of DDT. In: *DDT, the Insecticide Dichlorodiphenyltrichloroethane and Its Significance* (Paul Müller, ed.), Vol. II, pp. 11–251. Birkhäuser, Basel.

HAYES, W. J., JR., DALE, W. E., and LEBRETON, R. (1963). Storage of insecticides in French people. *Nature* **199**, 1189–1191.

LEWIN, J., and ESPLIN, D. W. (1961). Analysis of the spinal excitatory action of pentylenetetrazol. *J. Pharmacol. Exptl. Therap.* **132**, 245–250.

SHANES, A. M. (1951). Electrical phenomena in nerve. III. Frog sciatic nerve. *J. Cellular Comp. Physiol.* **38**, 17–40.

SHANKLAND, D. L. (1964). Involvement of spinal cord and peripheral nerves in DDT-poisoning syndrome in albino rats. *Toxicol. Appl. Pharmacol.* **6**, 197–213.

SUNG, C., and CHOW, Y. (1966). The biphasic effect of DDT and gamma-hexachlorocyclohexane on the biotransformation of pentobarbital *in vivo. Yao Hsueh Pao (Acta Pharmaceutica Sinica)* **13**, 119–125.

TIMIRAS, P. S., WOODBURY, D. M., and BAKER, D. H. (1956). Effect of hydrocortisone acetate, desoxycorticosterone acetate, insulin, glucagon and dextrose, alone or in combination, on experimental convulsions and carbohydrate metabolism. *Arch. Intern. Pharmacodyn. Therap.* **105**, 450–467.

WOODBURY, L. A., and DAVENPORT, V. D. (1952). Design and use of a new electroshock seizure apparatus, and analysis of factors altering seizure threshold and pattern. *Arch. Intern. Pharmacodyn. Therap.* **92**, 97–107.

WOOLLEY, D. E., and BARRON, B. A. (1968). Effects of DDT on brain electrical activity in awake, unrestrained rats. *Toxicol. Appl. Pharmacol.* **12**, 440–454.

WOOLLEY, D. E., and RUNNELLS, A. L. (1967). Distribution of DDT in brain and spinal cord of the rat. *Toxicol. Appl. Pharmacol.* **11**, 389–395.

WOOLLEY, D. E., BARRON, B. A., SEEMAN, D. J., and TIMIRAS, P. S. (1965). Alterations in rhinencephalic electrical activity after administration of pentobarbital and pentylenetetrazol. *Proc. West. Pharmacol. Soc.* **8**, 4–6.

Distribution of DDT, DDD, and DDE in Tissues of Neonatal Rats and in Milk and Other Tissues of Mother Rats Chronically Exposed to DDT

DOROTHY E. WOOLLEY AND GLORIA M. TALENS

It is well known that DDT crosses the placental barrier to enter fetal tissues during gestation in several mammalian species, including the dog (Finnegan *et al.*, 1949), rabbit (Pillmore *et al.*, 1963), mouse (Bäckström *et al.*, 1965), and man (Wasserman *et al.*, 1967; Fiserova-Bergerova *et al.*, 1967; Zavon *et al.*, 1969). DDT is also known to be excreted in milk, as has been demonstrated in the cow (see Laben, 1968, for references), man (Laug *et al.*, 1951), dog (Woodard *et al.*, 1945), and rat (Telford and Guthrie, 1945; Ottoboni and Ferguson, 1969). Furthermore, in the rat, exposure of the dam to high dietary levels of DDT (e.g., 100 and 200 μg/g) during pregnancy and lactation resulted in delayed development of the nervous system in the offspring, as shown by significant delays in the ages at which the startle and righting reflexes could be elicited (Woolley, 1970). Other work (unpublished data) showed that exposure of the dam during the lactational period only was sufficient to delay nervous system development in the off-spring, whereas exposure only during gestation had little or no effect. On the other hand, Al-Hachim and Fink (1967, 1968) found that a single administration of DDT during the second or third trimester of pregnancy in mice was sufficient to delay maturation of the nervous system as measured by delays in the acquisition of conditioned avoidance responses and audiogenic seizure responses. These findings indicate that under certain conditions DDT in the diet of the mother may adversely influence the development of the fetus or the neonate.

The present work was undertaken to investigate further the distribution of DDT and its metabolites between maternal and fetal or neonatal tissues, especially in order to compare the relative quantities retained by the offspring during gestation versus lactation. To do this, female rats were chronically exposed to several dietary levels of DDT—25, 100, and 200 μg/g—during pregnancy and the postparturitional lactational period. Newborn rats were sacrificed for analysis immediately after birth before suckling had occurred, so that the DDT and its metabolites in the neonatal tissues could only have have come from the dam during gestation. DDT levels in these tissues were compared with levels in littermates sacrificed about 10 hr after birth when suckling had occurred and the stomachs were full of milk. The milk in the stomachs was analyzed as an indication of the level of DDT in the milk. However, because it was likely that this milk had become concentrated during the process of absorption, the dams were milked, and DDT levels also determined in these samples. In addition, mother rats were sacrificed at weaning, in order to determine the levels of DDT in plasma and other tissues.

METHODS

Distribution of ^{3}H-DDT. In a preliminary study, 1 mCi/kg of ^{3}H-DDT[1] [1,1,1-trichloro-2,2-bis(p-chlorophenyl)ethane] was injected into the jugular vein of 2 anesthetized Sprague-Dawley rats on the 20th day of gestation. The rats weighed about 350 g. After 3 hr, samples of maternal blood, uterus and placenta and fetal brain, liver, kidneys and remaining carcass were collected from several fetuses for determining the levels of radioactivity. (The remaining fetuses and brain of the dam were used for autoradiographic localization of the ^{3}H-DDT, to be described elsewhere.) Radioactivity was extracted by sonifying the tissues in a 20-fold volume of toluene scintillator fluid[2] and centrifuging the homogenates for 15 min at 3000 g and 3°C. Radioactivity was determined with a Packard Tri-Carb Liquid Scintillation spectrometer. Tritiated toluene was added to each sample as an internal standard and the sample counts were corrected for machine efficiency and sample quenching.

Chronic exposure to DDT. DDT[3] was purified according to the method of Cook and Cook (1946) and the recrystallized p,p'-DDT was determined to be 99.9 % pure by gas chromatographic analysis and by melting point. DDD [1,1-dichloro-2,2-bis(p-chloro-phenyl)ethane] and DDE [1,1-dichloro-2,2-bis(p-chlorophenyl)ethylene][4] were obtained in purified form for use as gas chromatographic standards.

[1] 32 mCi/mmole, New England Nuclear, Boston, Massachusetts.
[2] Omnifluor, New England Nuclear, Boston, Massachusetts.
[3] Nutritional Biochemicals Corp., Cleveland, Ohio.
[4] Beckman Instruments Inc., Fullerton, California.

Adult 200–250 g Sprague-Dawley female rats were fed mash diets containing 25, 100, or 200 μg/g DDT or a control mash diet with about 0.1 μg/g DDT for several days, then bred with male rats not previously exposed to DDT. The DDT and control diets were maintained throughout pregnancy and lactation. DDT was incorporated into the diet by dissolving purified p,p'-DDT in acetone and then thoroughly mixing the acetone solution with a portion of mash. The moistened mash was then mixed for 8 hr in a mixer,[5] with an appropriate amount of untreated mash to make a final concentration of 200 μg DDT per gram mash. The mixed mash was spread out to provide a wide surface area and dried for at least 24 hr before use to make certain that no acetone remained. Portions of the 200 μg/g DDT mash were mixed for 8 hr with appropriate quantities of untreated mash to make the 25 and 100 μg/g DDT mash diets. Analyses of aliquots of the various diets showed that the diets were uniformly mixed with no loss of DDT during mixing.

Newborn rats were collected from each litter at 0–1 hr after birth, before suckling had occurred, and at about 10 hr after birth, when suckling had occurred and stomachs were full of milk. Usually 2 rats per time period from each of 6–8 litters per group were used for analyses. Six rats per litter were retained during the lactational period and were used for studies of development as reported in part elsewhere (Woolley, 1970; McDonald et al., 1970). Brain, liver, kidneys, stomach, and the rest of the rat were collected and analyzed for DDT and its two major metabolites, DDD and DDE.

Between the 5th and 15th days postpartum, milk samples were collected from the dams. Rats were injected with 2 USP units POP oxytocin[6] iv and with 20 mg/kg pentobarbital[7] ip, then milked after 30 min with a hand milker. The milker was a glass tube fitted with a small suction cup and a side arm connected to a vacuum line. Pulsations in the degree of suction were produced by manually covering and uncovering an opening in the side arm. Milk samples were weighed and frozen until analyzed.

At weaning, the dams were sacrificed by decapitation. Cerebellum, cerebral cortex, brain stem, remaining brain, spinal cord, liver, kidney, fat, and plasma were collected, weighed, and frozen until analyzed.

Tissue Analyses. DDT, DDD, DDE, and lipids in tissues of neonatal and adult rats chronically exposed to DDT were extracted by homogenizing tissues in a 20-fold volume of chloroform–methanol–chlorobenzene (2:1:1) with a Sonifier Cell Disruptor.[8] Samples difficult to sonify, such as the remaining carcass of newborn rats, were first homogenized using a Virtis "23" homogenizer[9] before sonifying. Homogenized tissues were filtered, then 0.2 volume of water was added to the filtrate, and after thorough mixing, the two layers were allowed to separate. The upper layer was siphoned off. The lower layer was washed four times with a wash solution made of the upper phase of a chloroform–methanol–chlorobenzene–water (2:1:1:1) mixture. Methanol was added to clear any emulsion. The extracting solvent mixture was evaporated on a flash evaporator and the sample rinsed several times with hexane, then evaporated to dryness and placed in a 45°C vacuum oven overnight. A Florisil column was prepared for the gas chromatographic clean-up procedure, as described by Beckman et al. (1966). The

[5] Twin shell dry blender, Patterson-Kelley Co., Inc., East Stroudsburg, Pennsylvania.
[6] Armour-Baldwin Laboratories, Omaha, Nebraska.
[7] Diabutal, Diamond Laboratories, Des Moines, Iowa.
[8] Model 185D, Branson Sonic Power, Danbury, Connecticut.
[9] Virtis Company, Inc., Gardiner, New York.

flask containing the dried sample was weighed. Then the sample was transferred from the flask onto the Florisil column with 10 ml of hexane followed by four 10-ml hexane rinses, and the column was eluted with 200 ml benzene. The flask was dried and re-weighed in order to determine tissue lipids gravimetrically. All solvents were double distilled. The gas chromatograph[10] was equipped with a tritium foil electron-capture detector, a 6 ft by $\frac{1}{8}$ in outside diameter glass column packed with 10 % DC 200 on 80/100 mesh Gas Chrom Q. Column temperature was 200°C and N_2 flow rate was 40 ml/min. Under these conditions column retention times were 5.6 min for DDE, 7.1 min for DDD, and 9.4 min for DDT, so that a clear separation of the 3 compounds was achieved (see Fig. 1). A strip chart recorder[11] was used to read out the gas chromatograph. The quantity of compound present was determined by integrating the area under each peak with a planimeter and comparing with the areas for the standards. When the concentra-tions of DDT, DDD, and DDE in a sample differed greatly, an aliquot of the tissue extract was run on the gas chromatograph at one gain setting to determine the concen-tration of the major constituent(s). Then the remaining extract was concentrated by evaporation and another aliquot analyzed at the same or greater gain. Recovery of DDT from 10 CNS and liver samples was 98–100% using these procedures.

RESULTS

Distribution of ^{3}H-DDT. The results of the preliminary study, carried out to deter-mine directly whether or not ^{3}H-DDT passed from maternal plasma to fetal tissues, showed that radioactivity per unit fresh tissue weight was higher in fetal liver than in maternal plasma, uterus, and placenta (Table 1). Radioactivity levels were at least twice

TABLE 1

DISTRIBUTION OF RADIOACTIVITY IN MATERNAL AND FETAL TISSUES
3 HOURS AFTER INJECTION OF ^{3}H-DDT[a]

Tissue	Sample (no.)	Radioactivity (dpm/μl or mg fresh tissue, $\bar{x} \pm$ SE)	Concentration (μg/g or ml fresh tissue, $\bar{x} \pm$ SE)
Maternal			
Plasma	2	391 and 390	0.86 and 0.86
Uterus	6	327 ± 15	0.72 ± 0.03
Placenta	15	737 ± 30	1.62 ± 0.06
Placental Buttons	6	651 ± 37	1.43 ± 0.08
Fetal			
Liver	6	1,197 ± 90	2.63 ± 0.18
Kidneys	6	539 ± 28	1.18 ± 0.06
Brain	6	356 ± 12	0.78 ± 0.03
Rest of Fetus	6	428 ± 20	0.94 ± 0.04

[a] Two rats were injected iv on day 20 of gestation with 1 mCi/kg or about 0.35 mCi/rat of ^{3}H-DDT. Placenta and uterus were cut into portions and these were analyzed separately. Placental buttons repre-sent both maternal and fetal portions of the placenta. Radioactivity has been converted to DDT con-centrations in the last column by assuming that all of the radioactivity was due to ^{3}H-DDT.

[10] Model 1200, Varian Aerograph, Walnut Creek, California.
[11] Speedomax W, Leeds-Northrup, Philadelphia, Pennsylvania.

as high in fetal liver as in other fetal tissues examined, i.e., kidneys, brain and the rest of the fetus. The high liver radioactivity levels no doubt resulted in part because of the arrangement of the fetal circulatory system, whereby blood from the placenta first perfused the fetal liver before entering the fetal systemic circulation. Of the maternal tissues, radioactivity levels were significantly higher in placenta than in uterus and plasma. Although no attempt was made in this study to identify the source of radioactivity, work on the rate of metabolism of DDT in young and adult rats (Henderson and Woolley, 1969) and on the distribution of DDT, DDD, and DDE in newborn rats (see below) argue that the major source of radioactivity was ^{3}H-DDT.

Chronic exposure to DDT. The distribution of DDT, DDD, and DDE in neonatal and maternal tissues during chronic exposure of the dam to 25, 100, or 200 μg/g DDT in the diet of the dam is shown in Tables 2–4. In newborn rats within an hour after birth, DDT concentrations in the whole rat ranged from 1.3 to 2.2 μg/g fresh tissue, DDD ranged from 0.1 to 1.0 μg/g, and DDE ranged from 0.3 to 1.0 μg/g in the 3 DDT exposure groups (Table 2). Thus, although there was an 8-fold range in DDT levels in

TABLE 2

TISSUE CONCENTRATIONS OF DDT, DDD, AND DDE IN NEONATAL RATS
BORN TO DAMS ON DIFFERENT LEVELS OF DIETARY DDT[a]

Tissue	p,p'-DDT		p,p'-DDD		p,p'-DDE	
	Newborn	10 hr	Newborn	10 hr	Newborn	10 hr
		25 μg/g DDT in diet of dam				
Brain	0.9 ± 0.2	1.2 ± 0.2	Trace	Trace	0.2 ± 0.0	0.3 ± 0.1
Liver	1.3 ± 0.2	6.1 ± 0.8	0.3 ± 0.0	0.5 ± 0.2	0.3 ± 0.0	1.0 ± 0.1
Kidneys	2.4 ± 0.7	2.7 ± 0.4	Trace	Trace	0.4 ± 0.1	0.6 ± 0.1
Stomach + contents	2.5 ± 0.9	51.8 ± 8.6	Trace	Trace	0.8 ± 0.3	11.3 ± 2.3
Rest of rat	1.1 ± 0.2	6.3 ± 0.7	0.1 ± 0.0	0.4 ± 0.1	0.2 ± 0.0	1.4 ± 0.2
Whole rat	1.3	8.1	0.1	0.4	0.3	2.0
		100 μg/g DDT in diet of dam				
Brain	1.5 ± 0.1	1.9 ± 0.1	Trace	Trace	0.3 ± 0.1	0.5 ± 0.1
Liver	2.9 ± 0.3	14.9 ± 2.9	0.6 ± 0.1	2.8 ± 0.8	0.9 ± 0.2	2.8 ± 0.4
Kidneys	2.2 ± 0.3	3.1 ± 0.5	Trace	Trace	0.6 ± 0.1	0.7 ± 0.1
Stomach + contents	3.2 ± 0.6	135.1 ± 8.9	Trace	Trace	1.1 ± 0.3	20.9 ± 2.5
Rest of rat	1.9 ± 0.1	6.1 ± 1.4	0.3 ± 0.0	1.1 ± 0.3	0.5 ± 0.0	1.5 ± 0.3
Whole rat	2.1	12.5	0.3	1.0	0.5	2.5
		200 μg/g DDT in diet of dam				
Brain	2.5 ± 0.2	4.1 ± 0.3	0.1 ± 0.1	0.1 ± 0.1	0.7 ± 0.1	2.5 ± 0.2
Liver	4.3 ± 0.3	29.1 ± 3.1	1.6 ± 0.3	3.7 ± 0.7	1.6 ± 0.2	7.7 ± 1.2
Kidneys	3.7 ± 0.4	8.3 ± 1.3	0.3 ± 0.1	0.4 ± 0.2	2.0 ± 0.2	6.3 ± 1.4
Stomach + contents	2.3 ± 0.2	237.7 ± 22.0	0.2 ± 0.1	4.4 ± 1.3	1.5 ± 0.3	54.1 ± 5.8
Rest of rat	2.0 ± 0.4	6.5 ± 1.1	0.8 ± 0.2	3.1 ± 0.3	1.0 ± 0.1	2.2 ± 0.2
Whole rat	2.2	22.3	1.0	3.4	1.0	5.5

[a] Newborn rats were removed from the dam 0–1 hr after birth and before suckling; others were removed about 10 hr after birth when stomachs were full of milk. Values are means ± SE for concentrations of DDT, DDD, or DDE expressed as μg/g of fresh tissue weight based on 6–20 samples of each tissue in each group. Average values for the whole rat were calculated from the concentration and weight data of constituent tissues.

TABLE 3

CONCENTRATIONS OF DDT, DDD, AND DDE IN MILK OF LACTATING MOTHER
RATS FED VARIOUS LEVELS OF DIETARY DDT[a]

DDT in diet (μg/g)	p,p'-DDT	p,p'-DDD	p,p'-DDE
25	32.7 ± 2.8	0.7 ± 0.1	2.9 ± 0.3
100	92.4 ± 4.6	1.1 ± 0.3	5.6 ± 0.9
200	174.0 ± 11.2	2.1 ± 1.0	20.2 ± 4.2

[a] Values are means ± SE for seven samples of fresh milk per group expressed as μg/g fresh milk. Percent extractable lipid in milk samples was 13.7 ± 0.5.

TABLE 4

CONCENTRATIONS OF DDT, DDD, AND DDE IN TISSUES OF LACTATING
MOTHER RATS ON DIETARY DDT[a]

Tissue	p,p'-DDT	p,p'-DDD	p,p'-DDE
		25 μg/g DDT in diet	
Cerebellum	2.0 ± 0.4	Trace	0.4 ± 0.1
Cerebral cortex	1.8 ± 0.4	Trace	0.4 ± 0.1
Brain stem	2.6 ± 0.4	Trace	0.5 ± 0.1
Remaining brain	2.1 ± 0.4	Trace	0.7 ± 0.3
Spinal cord	2.3 ± 0.4	Trace	0.6 ± 0.1
Liver	1.7 ± 0.2	2.5 ± 0.7	0.7 ± 0.2
Kidney	1.9 ± 0.2	0.1 ± 0.0	0.5 ± 0.1
Fat	129.8 ± 19.0	Trace	15.6 ± 3.1
Plasma	0.7 ± 0.2	Trace	0.3 ± 0.1
		100 μg/g DDT in diet	
Cerebellum	3.1 ± 0.3	0.1 ± 0.0	0.4 ± 0.0
Cerebral cortex	2.8 ± 0.3	Trace	0.4 ± 0.0
Brain stem	3.6 ± 0.2	0.1 ± 0.0	0.5 ± 0.1
Remaining brain	5.2 ± 0.8	0.2 ± 0.0	1.2 ± 0.0
Spinal cord	5.3 ± 0.6	0.1 ± 0.0	0.7 ± 0.1
Liver	5.1 ± 0.8	5.8 ± 0.4	1.0 ± 0.2
Kidney	4.7 ± 0.8	0.4 ± 0.1	0.6 ± 0.1
Fat	1,021.3 ± 165.4	12.8 ± 4.6	95.2 ± 15.3
Plasma	1.4 ± 0.3	Trace	0.9 ± 0.2
		200 μg/g DDT in diet	
Cerebellum	5.8 ± 0.3	0.2 ± 0.1	0.7 ± 0.1
Cerebral cortex	5.3 ± 0.3	0.1 ± 0.0	0.8 ± 0.1
Brain stem	7.1 ± 0.4	0.2 ± 0.1	0.9 ± 0.1
Remaining brain	5.6 ± 0.3	0.1 ± 0.0	0.7 ± 0.0
Spinal cord	5.2 ± 0.6	0.2 ± 0.0	1.2 ± 0.4
Liver	7.5 ± 1.1	7.9 ± 1.1	1.7 ± 0.2
Kidney	6.6 ± 1.3	0.3 ± 0.1	0.7 ± 0.2
Fat	1,346.1 ± 131.4	12.6 ± 1.0	123.7 ± 7.7
Plasma	1.7 ± 0.4	0.1 ± 0.0	0.3 ± 0.0

[a] Values are means ± SE expressed as μg/g fresh tissue weight or ml plasma for 6–8 samples per group. Rats were exposed to dietary DDT during the 3 wk of gestation and 3 wk of postparturition lactation.

the diet of the dam, the range of DDT and DDE concentrations in tissues of the offspring was only about 2- or 3-fold. The range and variation in DDD levels were greater, probably because DDD levels were relatively low and approached the limits of detectability with the methods used. In most tissues, concentrations of DDT were higher than concentrations of DDE, which in turn were higher than concentrations of DDD. Concentrations of DDT, DDD, and DDE were higher in liver than in brain in all DDT treatment groups.

About 10 hr after birth when the rats had suckled and the stomachs were full of milk, concentrations of DDT in the stomach and contents were increased 20- to 100-fold as compared with concentrations in neonates 1-hr after birth (Table 2). DDT concentrations in the stomach and contents were higher than in milk samples obtained by milking the dams (Table 3) undoubtedly because the milk in the stomachs had become concentrated by preferential absorption of water. Absorption of DDT and its metabolites from the milk in the stomachs was rapid, as shown by 7- to 20-fold increases in levels of DDT in the livers of rats 10 hr after birth, as compared with concentrations 1 hr after birth. The greatest increase in concentrations of DDT during the 10 hr after birth occurred in the stomach and its contents, as would be expected. The next greatest increase occurred in the liver, followed by the kidneys and the rest of the rat, and the smallest increase was in the brain.

Concentrations of DDT in milk samples obtained by milking the dams (Table 3) were 33 μg/g fresh milk from dams on 25 μg/g dietary DDT, 92 μg/g in milk from dams on 100 μg/g DDT, and 174 μg/g in milk from mothers on 200 μg/g DDT. Thus, at the lower level of dietary DDT, DDT levels in fresh milk were higher than concentrations in the diet, whereas the DDT output in milk was relatively lower than the dietary DDT levels in the dams on 100 and 200 μg/g DDT. Levels of DDD were almost insignificant in the milk, reaching only 2 μg/g even in the highest DDT treatment group, or about 1% of the concentration of DDT. DDE concentrations were about 10% of DDT concentrations and were 4–10 times higher than DDD concentrations in milk.

DDT concentrations in the plasma of the mother rats (Table 4) ranged from 0.7 to 1.7 μg/g fresh plasma for the 3 dietary levels of DDT, so that plasma levels differed only 2- to 3-fold. The range for DDT levels in the plasma of the dams was about the same or slightly lower than the range in tissue levels of DDT for the offspring immediately after birth (Tables 1, 2, and 4).

DDT levels in various tissues of the mother rats were higher than levels in corresponding tissues of the offspring immediately after birth (Tables 2 and 4). For example, in the 200 μg/g DDT treatment group, average DDT concentrations in brain areas of the dam ranged from 5 to 7 μg/g fresh tissue, whereas DDT concentration in the brain of the newborn rat was 2.6 μg/g. Also, in the same DDT exposure group, DDT concentrations were 7.5 μg/g in maternal liver and 4.3 μg/g in neonatal liver.

Of all of the tissues analyzed from the dam, fat contained by far the highest concentrations of DDT, DDD, and DDE (Table 4), as has been reported previously (Finnegan et al., 1949). Samples of fat also contained 88% extractable lipid, compared with 5% extractable lipid in liver.

The lipid levels of brain stem and spinal cord are more than twice as high as the levels in cerebral cortex and cerebellum, because of the high myelin content of the former (Woolley and Runnells, 1967). The fact that DDT concentrations per unit fresh tissue weight did not differ markedly among the various neural areas (Table 4) suggests that DDT is not particularly soluble in myelin; otherwise it would have distributed more in proportion to the myelin present.

Liver was the only tissue in which DDD concentrations were as high as or higher than DDT concentrations (Table 4, Fig. 1). DDE was present in much lower levels than DDD in the liver, whereas the reverse was true for all other tissues.

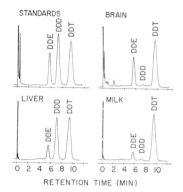

FIG. 1. Gas chromatograph charts showing the retention times and clear separation of DDT, DDD, and DDE in several tissue extracts and a solution of standards.

DISCUSSION

The present results show that in the rat DDT readily passes the placental barrier to enter tissues of the developing fetus, thus confirming and extending findings in other mammalian species as mentioned in the introduction. Even so, when the dam was chronically exposed to DDT during gestation, DDT concentrations in tissues of the offspring immediately after birth were relatively low, especially when compared with the same maternal tissues. The higher DDT tissue concentrations in the dam than in the neonate may be related to the higher lipid levels of tissues in adult rats as compared with lipid levels in tissues of immature rats and to the high oil:water partition coefficient of DDT, as suggested previously (Henderson and Woolley, 1969).

The concentrations of DDT in tissues of the offspring were relatively greater during the lactational period than during gestation. This was shown by the increased tissue concentrations of DDT in neonatal rats after suckling had occurred and the stomachs were full of milk, as compared with concentrations in newborn rats before suckling. The DDT levels in the offspring after suckling were very high, at least in part because of the high levels of DDT in the milk. By contrast, the DDT levels in maternal plasma, which was the source of fetal plasma levels of DDT, were relatively low. The DDT output in the milk of the dams was almost as high as the level in their diet when the DDT intake level was 100 or 200 μg/g diet and was even higher than the intake level at 25 μg/g dietary DDT. However, even 25 μg/g dietary DDT is much higher than dietary concentrations actually likely to be encountered, so it may be expected that at "normal" dietary levels of DDT, DDT concentrations in milk would be higher than in the diet. Similar conclusions were reached by Ottoboni and Ferguson (1969).

Two-way placental transport of dieldrin has been shown to occur in the rabbit (Hathaway et al., 1967), and probably also occurs with DDT. Hence, an equilibrium in distribution of DDT between fetal and maternal tissues may be a factor in promoting low levels of DDT in fetal tissues. This equilibrium would not occur after birth, of course.

It must be assumed that the DDT intake per unit body weight is higher in the preweanling rat than in the dam. This conclusion is based on the well known fact that food intake per unit body weight is greater in immature than in adult animals, and on the present observation that DDT levels in the milk are as high as levels in the diet of the dam.

The different ratios of DDD to DDE concentrations for liver versus other tissues may be due to the different half-lives and patterns of metabolism for the two compounds. The conversion of DDT to DDD or to DDE has been postulated to proceed along separate routes in rat (Peterson and Robison, 1964; see review by Hayes, 1965) and chicken (Abou-Donia and Menzel, 1968). A controversy over whether DDT was converted to DDD by the liver or by microorganisms in the gastrointestinal tract has apparently been resolved by the demonstration that DDD levels in liver were the same in conventional and microbe-free rats after feeding DDT, so that the contribution by intestinal microflora to tissue DDD levels appeared to be negligible (Ottoboni et al., 1968). Furthermore, the isolated perfused liver of the rat has been shown to convert DDT to DDD, which in turn is converted to still other metabolites; the kidney cannot metabolize either DDT or DDD (Peterson and Robison, 1964; Datta and Nelson, 1970). Hence, the high levels of DDD in the liver are due to the fact that it is produced by this organ, perhaps exclusively. The low levels of DDD in other tissues may be explained if only small quantities of DDD pass out of the liver into the circulation to reach other tissues, or if DDD is eliminated from the circulation relatively fast. In the pigeon, DDE was eliminated only slowly from all tissues (half-life 250 days), whereas DDD was eliminated more rapidly (half-life 24 days) (Bailey et al., 1969a,b). A similar difference in turnover for the two compounds may also exist in the rat.

ACKNOWLEDGMENTS

This work was supported by NIH grant ES-00163, and aided in part by the Health Sciences Advancement Award in Comparative Medicine to the University of California, Davis.

REFERENCES

ABOU-DONIA, M. B., and MENZEL, D. B. (1968). The metabolism in vivo of 1,1,1-trichloro-2,2-bis(p-chlorophenyl)ethane (DDT), 1,1-dichloro-2,2-bis(p-chlorophenyl)ethane (DDD), and 1,1-dichloro-2,2-bis(p-chlorophenyl)ethylene (DDE) in the chick by embryonic injection and dietary ingestion. Biochem. Pharmacol. 17, 2143–2161.

AL-HACHIM, G. M., and FINK, G. B. (1967). Effect of DDT or parathion on the audiogenic seizure of offspring from DDT or parathion-treated mothers. Psychol. Rep. 20, 1183–1187.

AL-HACHIM, G. M., and FINK, G. B. (1968). Effect of DDT or parathion on condition avoidance response of offspring from DDT or parathion-treated mothers. Psychopharmacologia 12, 424–427.

BÄCKSTRÖM, J., HANSSON, E., and ULLBERG, S. (1965). Distribution of C14-DDT and C14-dieldrin in pregnant mice determined by whole-body autoradiography. Toxicol. Appl. Pharmacol. 7, 90–96.

BAILEY, S., BUNYAN, P. J., RENNISON, B. D., and TAYLOR, A. (1969a). The metabolism of 1,1-di(p-chlorophenyl)-2,2,2-trichloroethane and 1,1-di(p-chlorophenyl)-2,2-dichloroethane in the pigeon. Toxicol. Appl. Pharmacol. 14, 13–22.

BAILEY, S., BUNYAN, P. J., RENNISON, B. D., and TAYLOR, A. (1969b). The metabolism of 1,1-di(p-chlorophenyl)-2,2-dichloroethylene and 1,1-di(p-chlorophenyl)-2-chloroethylene in the pigeon. *Toxicol. Appl. Pharmacol.* **14**, 23–32.

BECKMAN, H., BEVENUE, A., CARROLL, K., and ERRO, F. (1966). An improved method for analysis of DDT and its metabolites in eggs. *J. Ass. Offic. Anal. Chem.* **49**, 996–999.

COOK, K. H., and COOK, W. A. (1946). A simple purification procedure for DDT. *J. Amer. Chem. Soc.* **68**, 1663–1664.

DATTA, P. R., and NELSON, M. J. (1970). p,p'-DDT detoxication by isolated perfused rat liver and kidney. In: *Pesticides Symposia* (W. B. Deichmann, ed.). Halos and Associates, Miami, Florida.

FINNEGAN, J. K., HAAG, H. B., and LARSON, P. S. (1949). Tissue distribution and elimination of DDD and DDT following oral administration to dogs and rats. *Proc. Soc. Exp. Biol. Med.* **72**, 357–360.

FISEROVA-BERGEROVA, V., RADOMSKI, J. L., DAVIES, J. E., and DAVIS, J. H. (1967). Levels of chlorinated hydrocarbon pesticides in human tissues. *Ind. Med. Surg.* **36**, 65–70.

HATHAWAY, D. E., MOSS, J. A., ROSE, J. A., and WILLIAMS, D. J. M. (1967). Transport of dieldrin from mother to blastocyst and from mother to fetus in pregnant rabbits. *Eur. J. Pharmacol.* **1**, 167–175.

HAYES, W. J., JR. (1965). Review of the metabolism of chlorinated hydrocarbon insecticides especially in mammals. *Annu. Rev. Pharmacol.* **5**, 27–52.

HENDERSON, G. L., and WOOLLEY, D. E. (1969). Studies on the relative insensitivity of the immature rat to the neurotoxic effects of 1,1,1-trichloro-2,2-bis(p-chlorophenyl)ethane (DDT). *J. Pharmacol. Exp. Ther.* **170**, 173–180.

LABEN, R. C. (1968). DDT contamination of feed and residues in milk. *J. Animal Sci.* **27**, 1643–1650.

LAUG, E. P., KUNZE, F. M., and PRICKETT, C. S. (1951). Occurrence of DDT in human fat and milk. *Arch. Ind. Hyg.* **3**, 245–246.

MCDONALD, L. W., WOOLLEY, D. E., and BARNES, P. (1970). The effects of DDT [1,1,1-trichloro-2,2-bis(p-chlorophenyl)ethane] on the developing nervous system of the rat. (Abstr.) *Proc. Int. Congr. Neuropathol. 6th*, pp. 105 and 130.

OTTOBONI, A., and FERGUSON, J. I. (1969). Excretion of DDT compounds in rat milk. *Toxicol. Appl. Pharmacol.* **15**, 56–61.

OTTOBONI, A., GEE, R., STANLEY, R. L., and GOETZ, M. E. (1968). Evidence for conversion of DDT to TDE in rat liver: II. Conversion of p,p'-DDT to p,p'-TDE of axenic rats. *Bull. Environ. Contam. Toxicol.* **3**, 302–308.

PETERSON, J. E., and ROBISON, W. H. (1964). Metabolic products of p,p'-DDT in the rat. *Toxicol. Appl. Pharmacol.* **6**, 321–327.

PILLMORE, R. E., KEITH, J. O., MCEWEN, L. C., MOHN, M. H., WILSON, R. A., and ISE, G. H. (1963). Cottontail rabbit: feeding test. *U.S. Fish Wildl. Serv. Circ.* **167**, 47–50.

TELFORD, H. S., and GUTHRIE, J. E. (1945). Transmission of the toxicity of DDT through the milk of white rats and goats. *Science* **102**, 647.

WASSERMAN, M., WASSERMAN, D., ZELLERMAYER, L., and GON, M. (1967). Pesticides in people. Storage of DDT in the people of Israel. *Pesticides Monitoring J.* **1**, 15–20.

WOODARD, G., OFNER, R. R., and MONTGOMERY, C. M. (1945). Accumulation of DDT in the body fat and its appearance in the milk of dogs. *Science* **102**, 177–178.

WOOLLEY, D. E. (1970). Effects of DDT on the nervous system of the rat. In: *The Biological Impact of Pesticides in the Environment* (J. W. Gillett, ed.), Environmental Sci. Ser. No. 1. Oregon State Univ. Press, Corvallis, Oregon. pp. 114–124, pp. 142–145.

WOOLLEY, D. E., and RUNNELLS, A. L. (1967). Distribution of DDT in brain and spinal cord of the rat. *Toxicol. Appl. Pharmacol.* **11**, 389–395.

ZAVON, M. R., TYE, R., and LATORRE, L. (1969). Chlorinated hydrocarbon insecticide content of the neonate. In: *Biological Effects of Pesticides in Mammalian Systems, Ann. N. Y. Acad. Sci.* **160**, 196–200.

67

The Ultrastructure of Livers of Rats Fed DDT and Dieldrin

Renate D. Kimbrough, MD;

Thomas B. Gaines; and

Ralph E. Linder

B OTH dieldrin (1,2,3,4,10,10-hexachloro-6,7 - epoxy - 1,4,4a,5,6,7,8,8a - octahydro - 1,4-*endo-exo* - 5,8 - dimethanonaphthalene) and DDT [1,1,1-trichloro-2, 2-bis(*p*-chlorophenyl) ethane], if ingested in sufficient quantity over a period of time, lead to enlargement of the liver in rats and dogs.[1-4] Both compounds stimulate microsomal enzymes and increase hepatic smooth endoplasmic reticulum.[5-8]

Since concomitant environmental exposure to DDT and dieldrin can occur, and since there are indications that dieldrin metabolism is affected by DDT,[9,10] the effect of DDT and dieldrin alone and in combination on the ultrastructure of the rat liver was studied. The hexobarbital sleeping time of rats treated with these compounds was also determined. The results of these studies are reported in this paper.

Materials and Methods

Ninety Sherman strain male rats, 3 to 4 months old and weighing 342 to 512 gm, were distributed according to a table of random numbers into six groups of 15 rats each. The rats were individually caged and weighed each week.

DDT and dieldrin were technical grade and were fed as a component of the diet. Dietary levels fed to the respective groups of rats are presented in Table 1. The appropriate amount of each compound to give the final dietary concentration desired was diluted in 10 gm of corn starch by grinding in a mortar and pestle. This mixture was then added to the appropriate amount of ground laboratory chow and mixed with a mechanical mixer. Rats were fed the respective diets for eight weeks; food consumption in each group was measured during the sixth week of treatment. The hexobarbital sleeping time was tested in ten rats of each group eight weeks after onset of exposure to the treated diets. The hexobarbital was dissolved in water, 125 mg/kg was injected intraperitoneally (ip) into each rat, and the time from the loss of the righting reflex to its recovery was measured. The remaining five rats of each group were killed and the livers were removed for microscopic study. The tissue was fixed in phosphate-buffered 10% formaldehyde solution and stained with hematoxylin-eosin for examination under the light microscope. Formaldehyde solution-fixed frozen sections were stained with oil red O to show fat accumulation. Tissue for the electron microscope was fixed in phosphate-buffered 5% glutaraldehyde, postfixed in phosphate-buffered 1% osmium tetraoxide, dehydrated, and embedded in resin (Maraglas). Sections were cut with a glass knife, stained with lead citrate,[11] and examined with an electron microscope (Philips 300).

In a preliminary study, five adult Sherman strain rats were fed a diet containing 500 ppm of technical DDT and 100 ppm of dieldrin for six weeks and their livers were studied with the light and the electron microscopes.

Results

The animals fed 500 ppm of DDT showed a light tremor and those in the preliminary study fed 500 ppm of DDT and 100 ppm of

68

Treatment	Parameter	Body Weight Gain, gm	Liver Weight, % of Body Weight	Sleeping Time, min
None (controls)	N	15	15	7
	Range	62 — 139	2.41 — 3.06	46 — 98
	Mean ± SE	93 ± 5.7	2.81 ± 0.05	67.1 ± 6.9
Dieldrin, 50 ppm	N	15	15	9
(2.64 mg/kg/day)	Range	36 — 103	2.51 — 3.70	10 — 58
	Mean ± SE	77.0 ± 4.6	3.25 ± 0.08*	39.3 ± 4.3§
Dieldrin, 100 ppm	N	15	15	9
(5.33 mg/kg/day)	Range	34 — 108	3.26 — 4.17	24 — 45
	Mean ± SE	73.0 ± 4.9	3.74 ± 0.07*†	35.7 ± 2.2§
DDT, 250 ppm	N	15	15	9
(12.93 mg/kg/day)	Range	55 — 111	3.09 — 4.35	5 — 47
	Mean ± SE	85.0 ± 4.6	3.47 ± 0.10*	31.6 ± 4.5§
DDT, 500 ppm	N	15	15	9
(26.5 mg/kg/day)	Range	13 — 111	3.07 — 4.05	15 — 62
	Mean ± SE	66.0 ± 6.9	3.60 ± 0.08*†	30.4 ± 4.5§
DDT, 250 ppm	N	15	15	8
(13.3 mg/kg/day), +	Range	24 — 116	3.49 — 5.55	20 — 40
dieldrin, 50 ppm	Mean ± SE	74.0 ± 6.6	4.27 ± 0.11*‡	32.5 ± 3.0§
(2.66 mg/kg/day)				
DDT, 500 ppm	N	5		
(25.3 mg/kg/day), +	Range	5 — 32		
dieldrin, 100 ppm	Mean ± SE	20.0 ± 5.9	...	...
(5.06 mg/kg/day)				

* Significantly heavier than the livers of control rats (P <0.05).
† Significantly heavier than the livers of rats fed 50 ppm dieldrin (P <0.005).
‡ Significantly heavier than the livers of rats fed DDT or dieldrin alone (P <0.01).
§ Sleeping time significantly shorter than that of the controls (P <0.05).

dieldrin had a moderate to severe tremor (Table 1). This last group also gained less weight than the others. The food consumption of all treated animals did not differ from that of the controls.

The sleeping time (Table 1) of all treated groups was significantly reduced from that of the controls ($P < 0.005$) but no significant difference was found among the various treated groups. However, all of the dosage levels used were large, and stimulation of microsomal enzymes and reduction of sleeping time may have been maximal at the lowest dose tested. An occasional animal among the controls as well as in the treated group was not affected at all by hexobarbital. These animals were not included in the statistical evaluation (Table 1).

The mean liver weight and the percent of body weight of the livers (Table 1) were significantly increased in the treated groups ($P < 0.05$). A significant difference in the liver weights was also found between the rats fed 50 ppm of dieldrin and those given 100 ppm of dieldrin or 500 ppm of DDT. The liver weights of the animals fed the combination of 50 ppm of dieldrin and 250 ppm of DDT were significantly higher than those of the animals fed DDT or dieldrin alone.

There was no significant difference between the liver weights of rats given 250 or 500 ppm of DDT.

The light and electron microscopic findings are summarized in Table 2. The lipid concentrations in the liver sections varied greatly; however, the variations could not be related to the dietary concentrations of the two chlorinated hydrocarbons. The lowest doses of dieldrin (50 ppm) and DDT (250 ppm) given alone caused enlargement of liver cells around the central veins. The cytoplasm had a smoother appearance and at the 50-ppm dieldrin level, margination was also observed. The dietary concentrations of 100 ppm of dieldrin and 500 ppm of DDT caused cytoplasmic inclusions in a number of cells and the enlargement of the liver cells had spread to a larger portion of the liver lobules. The combination of dieldrin at 50 ppm and DDT at 250 ppm resulted in the enlargement of cells throughout the entire liver lobules and a greater number of liver cells had inclusions in their cytoplasm (Fig 1 and 2). With the combination of dieldrin at 100 ppm and DDT at 500 ppm, inclusions were observed in the cytoplasm of almost all cells and the nuclei of many cells appeared to be exceptionally large. The cytoplasmic

Table 2.—Light and Electron Microscopic Findings in the Livers of Rats Treated With Dieldrin and DDT

Dietary Level	Light Microscopic Findings	Electron Microscopic Findings
Dieldrin, 50 ppm	Enlarged cells around central veins, smooth-looking cytoplasm with margination; lipids, none to a large amount	Moderate increase in SER, atypical mitochondria
Dieldrin, 100 ppm	Enlarged cells around central veins, smooth-looking cytoplasm with margination; inclusions in cytoplasm of 3 of 5 livers; lipids, slight to moderate	Moderate increase in SER with swelling in 2 livers; atypical mitochondria
DDT, 250 ppm	Cells around central veins slightly enlarged; smooth-looking cytoplasm; lipids, mild to moderate	Mild to moderate increase in SER, occasional large myelin figures; atypical mitochondria seen in 1 of 5 animals
DDT, 500 ppm	Enlarged cells except in the periphery of the lobules; moderate number of vacuolated cells with inclusions in 2 of 5 rats; margination; lipids, moderate in 2 of 5	Less glycogen in 2 of 5 livers; marked increase in SER, swollen in some areas; occasional myelin figures; occasional atypical mitochondria in 3 of 5 livers
Dieldrin, 50 ppm, + DDT, 250 ppm	All cells enlarged, smooth cytoplasm, margination, inclusions moderate to many; lipids, mild to moderate	Moderate to marked increase in SER, swollen in some areas; less glycogen, slight to moderate number of atypical mitochondria
Dieldrin, 100 ppm, + DDT, 500 ppm	All cells enlarged, many cells with exceptionally large nuclei; margination in cytoplasm and inclusions in almost all cells; lipids, not studied	Marked increase in SER, which is swollen; indistinct cell borders; atypical mitochondria; vacuolated areas in cytoplasm surrounded by layers of dense lamellated material; occasional large vacuoles in cytoplasm containing granular material
None	Liver cells normal; lipids, none to moderate	Occasional small myelin figures

inclusions have been described previously.[2,4,7] They differ in appearance from acidophil bodies which are observed, for instance, in viral hepatitis or from Councilman bodies observed in yellow fever. They are usually not homogeneous but are composed of concentric rings of acidophil cytoplasm which surround one or more vacuoles, and they are observed within the cytoplasm of the cells.

On examination of liver sections with the electron microscope (Fig 3 to 8), a number of changes were observed in the cells of the treated animals. An increase in smooth endoplasmic reticulum (Fig 4) was noted in all treated rats and was most pronounced in the rats fed DDT at 500 ppm and in rats fed the combinations. A widening (swelling) of the vesicles of the smooth endoplasmic reticulum was observed in the livers of some of these rats (Table 2).

Many of the livers of the treated animals showed morphological changes of some of the mitochondria (Fig 5 to 6). These changes consisted of an increase in the number of cristae, a double outer membrane or loss of a portion of the outer membrane, and confluence of two mitochondria. The effect on the mitochondria was more pronounced in the livers of rats fed dieldrin or the combi-

nations. Mitochondrial ring-like formations were also observed (Fig 6).

The glycogen appeared to be decreased in the livers of rats fed DDT at 500 ppm and in those of the rats fed the combinations. The number of free ribosomes varied a great deal and did not seem to be dose-dependent. The liver cells of all rats, including the controls, contained lipid vacuoles which varied greatly in number and size. Some of these vacuoles contained several layers of dark membranes in their outer aspects. In addition, myelin figures were observed in many of the livers of treated rats and occasionally also in the controls. Quantitative differences of these various cytoplasmic components are difficult to assess, since only very small portions of the liver are examined under the electron microscope.

The morphological appearance of the liver cells of rats fed the combination of DDT at 250 ppm and dieldrin at 50 ppm differed from that of rats fed DDT or dieldrin alone by its more pronounced increase in smooth endoplasmic reticulum. In this group, the smooth endoplasmic reticulum of all livers examined showed dilatation "swelling" of the vesicles in some areas of the cell, usually in the periphery. The livers of the rats fed the combination of 500 ppm of DDT and

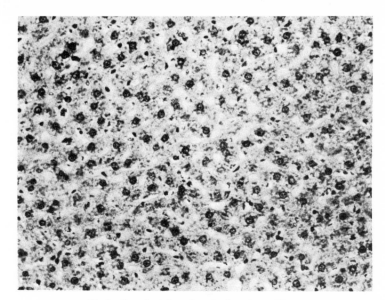

Fig 1.—Normal rat liver (hematoxylin-eosin, × 250).

Fig 2.—Section of liver from rat fed 50 ppm dieldrin and 250 ppm DDT. Note enlargement of liver cells, margination, and cytoplasmic inclusions **(arrows)** (hematoxylin-eosin, × 250).

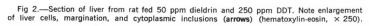

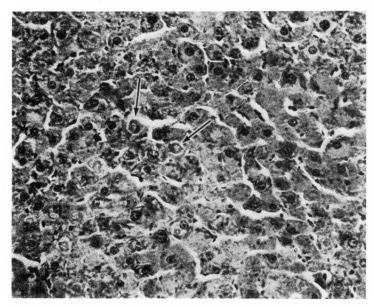

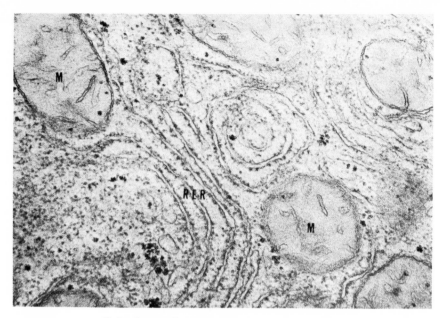

Fig 3.—Segment of a normal rat liver cell. **M** indicates mitochondria; and **RER,** rough endoplasmic reticulum (lead citrate, × 42,000).

Fig 4.—Portion of liver cell from a treated rat illustrating increase in smooth endoplasmic reticulum which appears swollen (lead citrate, × 80,000).

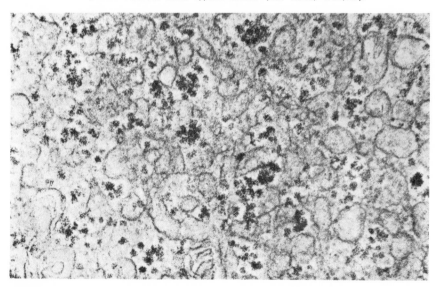

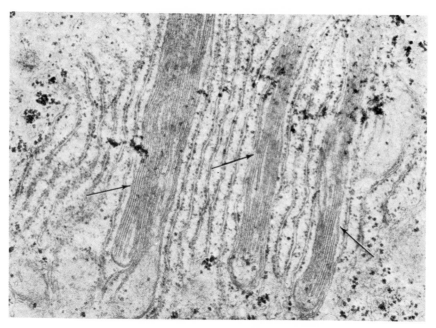

Fig 5.—Portion of liver cell from a treated rat showing an increased number and parallel arrangement of cristae within mitochondria (**arrows**) (lead citrate, × 64,000).

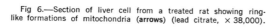

Fig 6.—Section of liver cell from a treated rat showing ring-like formations of mitochondria (**arrows**) (lead citrate, × 38,000).

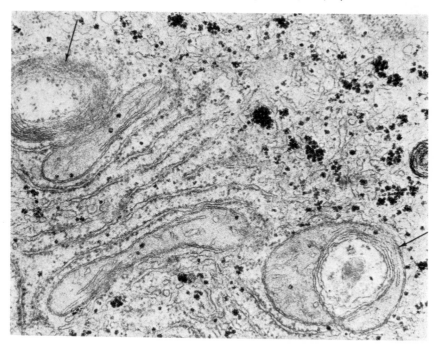

73

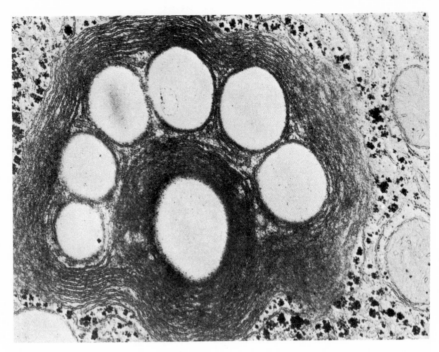

Fig 7.—Concentrically arranged membranes in the cytoplasm of a liver cell from a treated rat. The membranes surround lipid-containing vacuoles (lead citrate, × 24,000).

Fig 8.—Concentrically arranged membranes and an area of finely granular structureless material in the liver of a treated rat (lead citrate, × 27,000).

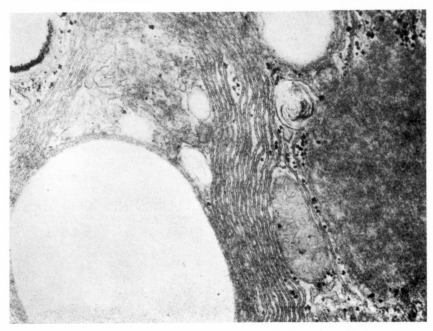

100 ppm of dieldrin also showed this effect on the smooth endoplasmic reticulum. In addition, these liver cells had indistinct cell borders and vacuolated areas in the cytoplasm which were surrounded by layers of concentrically arranged membranes (Fig 7). Occasionally large areas were observed which contained a granular material but no other cell organelles (Fig 8). The granular material may represent tangentially sectioned layers of concentrically arranged membranes.

Comment

The results of this study show that the hexobarbital sleeping time is reduced in rats pretreated for eight weeks with DDT or dieldrin. The reduction in sleeping time was similar in rats given different dosages of DDT or dieldrin or the combination of the two. However, the liver weights were higher in the animals given the combination of DDT and dieldrin. The livers of the rats given dieldrin at 100 ppm also weighed more than those given dieldrin at only 50 ppm. The increase in liver weight was associated with an increase in smooth endoplasmic reticulum demonstrated with the electron microscope. At present it is widely accepted that the increase in smooth endoplasmic reticulum corresponds to an increase in microsomal enzyme activity of the livers.[7,12-14] However, Hutterer et al[8] found that endoplasmic reticulum can become hypoactive. They gave rats daily small ip doses of dieldrin for 28 days and then increased the dose of dieldrin. The liver weight and the microsomal protein continued to increase but the activity of the drug handling enzymes decreased progressively. The point at which an increase in smooth endoplasmic reticulum is still noted while the microsomal enzyme activity remains stationary or declines will probably show a great species, sex, age, and time-dose related variation, when microsomal enzymes are evaluated for their ability to metabolize additional chemicals. This was also emphasized by Gillett and Arscott[15] in a recent article.

The combination of DDT and dieldrin produces a more pronounced increase in smooth endoplasmic reticulum than either compound alone. The combination also caus-es enlargement of a greater proportion of the hepatocytes, and a greater number of cells with inclusions in their cytoplasm are observed under the light microscope.

Estelle Gray assisted with the statistical evaluation and Mary Anne Dobbs assisted with the electron microscopic studies. Annie Alford prepared the tissue for the light microscopic examination.

Nonproprietary and Trade Names of Drug

Hexobarbital—*Evipal, Sombulex, Somnalert.*

References

1. Fitzhugh OG, Nelson AA: The chronic oral toxicity of DDT (2,2-bis-p-chlorophenyl-1,1,1-trichloroethane). *J Pharmacol Exp Ther* 89:18-30, 1947.
2. Ortega P, Hayes WJ, Durham WF, et al: *DDT in the Diet of the Rat,* monograph 43. Public Health Service, 1956.
3. Lehman AJ: *Summaries of Pesticide Toxicity.* Topeka, Kan, Association of Food and Drug Officials of the United States, 1965, pp 16-18, 19-21.
4. Kimbrough RD, Gaines TB, Hayes WJ: Combined effect of DDT, pyrethrum and piperonyl butoxide in rat liver. *Arch Environ Health* 16:333-341, 1968.
5. Hart LG, Fouts JR: Effects of acute and chronic DDT administration on hepatic microsomal drug metabolism in the rat. *Proc Soc Exp Biol Med* 114:388-392, 1963.
6. Ghazal A, Koransky WB, Portig T, et al: Beschleunigung von Entgiftungsreaktionen durch verschiedene Insekticide. *Arch Exp Path* 249:1-10, 1964.
7. Ortega P: Light and electron microscopy of dichlorodiphenyltrichloroethane (DDT) poisoning in the rat liver. *Lab Invest* 15:657-679, 1966.
8. Hutterer F, Schaffner F, Klion FM, et al: Hypertrophic, hypoactive smooth endoplasmic reticulum: A sensitive indicator of hepatotoxicity, exemplified by dieldrin. *Science* 161:1017-1019, 1968.
9. Street JC: DDT antagonism to dieldrin storage in adipose tissue of rats. *Science* 146:1580-1581, 1964.
10. Street JC: Organochlorine insecticides and the stimulation of liver microsome enzymes. *Ann NY Acad Sci* 160:274-290, 1969.
11. Reynolds ES: The use of lead citrate at high pH as an electron-opaque stain in electron microscopy. *J Cell Biol* 17:208-212, 1963.
12. Remmer H, Merker HJ: Drug induced changes in the liver endoplasmic reticulum: Association with drug metabolizing enzymes. *Science* 142:1657-1658, 1963.
13. Fouts JR, Rogers LA: Morphological changes in the liver accompanying stimulation of microsomal drug metabolizing enzyme activity by phenobarbital, chlordane, benzypyrene or methylchloranthrene in rats. *J Pharmacol Exp Ther* 147:112-119, 1965.
14. Greim H, Ueberberg H, Remmer H: Veränderungen des endoplasmatischen Retikulums durch Phenobarbital und DDT. *Arch Exp Path* 257:24-25, 1967.
15. Gillett JW, Arscott GH: Microsomal epoxidation in Japanese quail: Induction by dietary dieldrin. *Comp Biochem Physiol* 30:589-600, 1969.

METABOLIC CONTROL MECHANISMS IN MAMMALIAN SYSTEMS—IX

ESTROGEN-LIKE STIMULATION OF UTERINE ENZYMES BY o,p'-1,1,1-TRICHLORO-2-2-BIS (p-CHLOROPHENYL) ETHANE*

RADHEY L. SINGHAL, J. R. E. VALADARES and WAYNE S. SCHWARK

IT IS becoming increasingly apparent that several drugs as well as a variety of commonly used insecticides are capable of producing detrimental effects upon reproduction and fertility in several species. Several investigators have suggested that the chlorinated hydrocarbon, 1,1,1-trichloro-2-2-bis (p-chlorophenyl) ethane (DDT), may produce hormonal or antifertility effects in laboratory animals and in wildlife.[1-5] Treatment

* Supported in part by a grant from the Upjohn Company, the Medical Research Council of Canada and Eli Lilly & Co., Indianapolis, Ind. A preliminary account of portions of this work has appeared previously in *Proc. Fourth Int. Cong. Pharmac.*, Basel, Switzerland (1969).

of immature female or ovariectomized rats with various isomers of DDT results in estrogenic effects that are more pronounced with o,p'-DDT than with p,p'-DDT. Administration of o,p'-DDT enhances uterine water imbibition, wet weight and incorporation of glucose-^{14}C into uterine protein, lipid and RNA.[4] In addition, Bitman et al.[5] demonstrated that this isomer of DDT is capable of augmenting glycogen accumulation in uteri of immature rats as well as in the oviduct of two avian species.

The possible physiological and ecological implications of this pesticide on animal fertility, as well as the similarity in configuration of DDT to the synthetic estrogen diethylstilbesterol, prompted us to examine the influence of DDT on several important carbohydrate-metabolizing enzymes in the uterus of the ovariectomized rat. Our earlier studies have shown that diethylstilbesterol, as well as the naturally occurring estrogens, regulates uterine glucose metabolism by exerting its effects on receptor sites at the source of enzyme formation to turn on the biosynthesis of certain enzymes, *viz.* hexokinase (HK), phosphofructokinase (PFK), aldolase and pyruvate kinase (PK), that are involved in the process of glycolysis.[6-12] In the present communication, we report that o,p'-DDT is capable of exerting an estrogen-like modulating influence on several glycolytic and hexosemonophosphate shunt enzymes in uteri of ovariectomized rats.

MATERIALS AND METHODS

Young female rats of the Wistar strain, weighing approximately 150 g at the time of surgery, were used in this study. The animals were ovariectomized bilaterally under light pentobarbital anesthesia and were used after a postoperative period of 2 weeks.[6, 7, 10] The following experimental procedures were used.

Comparison of the effects of two DDT analogs. The effects of o,p'-DDT and p,p'-DDT on uterine weight, glycogen content and the activities of several glycolytic and pentose shunt enzymes were studied in groups of ovariectomized rats. Single doses of 10 mg/100 g of each insecticide were administered i.m., and animals were sacrificed 16 hr later.

Time-course of o,p'-DDT-induced changes. Since the above study indicated that o,p'-DDT was the more potent isomer, the sequence of events after administration of this hydrocarbon was followed for a period of 24 hr. Groups of ovariectomized rats were given single i.m. injections of o,p'-DDT (10 mg/100 g) and killed after 4, 8, 16 and 24 hr.

Effects of actinomycin and cycloheximide. In order to investigate the nature of the o,p'-DDT-induced increases in uterine enzyme activities, the effects of two inhibitors of RNA and protein synthesis were examined. Four groups of ovariectomized animals were used: (1) control rats; (2) animals treated with o,p'-DDT; (3) and (4) o,p'-DDT-treated rats injected (i.p.) with either actinomycin or cycloheximide 30 min prior to administration of the insecticide. All animals were killed 16 hr after o,p'-DDT injection.

Influence of cycloheximide on chronic o,p'-DDT treatment. The effect of cyclo-heximide on uterine enzyme increases induced by treatment with o,p'-DDT for 3 days was studied in the following three groups of ovariectomized rats: (1) control rats; (2) animals injected with o,p'-DDT; (3) o,p'-DDT-treated rats given cycloheximide. o,p'-DDT (5 mg/100 g) was injected at 24-hr intervals for 3 days. Cycloheximide was administered daily 30 min prior to injection of the insecticide. Groups of rats were killed 24 hr after termination of o,p'-DDT treatment.

Effect of progesterone and estradiol-17β. In order to examine the effects of progesterone and estradiol-17β on o,p'-DDT-induced changes in uterine weight and enzyme activities, the following six groups of ovariectomized rats were employed: (1) control animals; (2, 3 and 4) rats treated with o,p'-DDT (10 mg/100 g), progesterone (5 mg/100 g) and estradiol-17β (0·1 μg/100 g) respectively; (5 and 6) o,p'-DDT-treated animals injected 30 min earlier with either progesterone or estradiol-17β. All animals were killed 16 hr after administration of the insecticide.

Chemicals and dosages. Analytically pure o,p'-DDT and p,p'-DDT were obtained from the Aldrich Chemical Co., suspended in corn oil and administered i.m., in doses of 10 mg/100 g body weight. Actinomycin D (25 μg/100 g; Merck) and cycloheximide (70 μg/100 g; Upjohn) were dissolved in 0·9% saline and given by the i.p. route. Estradiol-17β (0·1 μg/100 g; Sigma) and progesterone (5 mg/100 g; Nutritional Biochemicals) were dissolved in ethanolic 0·9% saline and administered i.m.

Sample preparation and assay methods. All animals were stunned, decapitated and exsanguinated. Uteri were rapidly excised, trimmed of all extraneous tissue and weighed on a Roller Smith torsion balance. Pooled uteri were finely minced in a beaker immersed in crushed ice, and 5% homogenates and supernatant fluids were prepared as described in previous communications.[6, 7] The activities of hexokinase,[11] phosphofructokinase,[6] aldolase[12] and pyruvate kinase[10, 13] were assayed in the supernatant. Glucose 6-phosphate dehydrogenase (G6-PDH) and 6-phosphogluconate dehydrogenase (6-PGDH) were estimated according to the method of Glock and McLean,[14] as described previously.[15, 16] Enzyme activities were assayed under strictly linear kinetic conditions at 340 mμ in a constant recording Unicam spectrophotometer, model SP 800, set at 37° and calculated as micromoles of substrate metabolized per gram of tissue per hour times the weight of the uterus.[6-12] Uterine glycogen was assayed by the anthrone method of Seifter *et al.*,[17] and expressed as micrograms per total uterus. The results were subjected to statistical evaluation and significant differences between the means (calculated as P values) are shown. No statistical significance is indicated when the P value is $>0·05$.

RESULTS AND DISCUSSION

Effects of DDT analogs on uterine weight, enzyme activities and glycogen content. Administration of o,p'-DDT (10 mg/100 g) to ovariectomized rats resulted in marked increases in uterine weight and in the activities of several uterine carbohydrate-metabolizing enzymes (Table 1). Sixteen hr after injection of o,p'-DDT, uterine wet weight increased to 175 per cent, while hexokinase, phosphofructokinase, aldolase and pyruvate kinase activities were elevated to more than 200 per cent of the control values. The activity of G6-PDH was increased to 285 per cent and that of 6-PGDH to 197 per cent in uteri of rats treated with o,p'-DDT. Administration of p,p'-DDT also resulted in significant increases in the activities of hexokinase, phosphofructokinase, aldolase, pyruvate kinase and G6-PDH, although these were of a much smaller magnitude than those observed with o,p'-DDT. Uterine weight and the activity of 6-PGDH remained relatively unaffected by p,p'-DDT treatment. The effects of o,p'-DDT and p,p'-DDT on uterine glycogen content are also shown in Table 1. Whereas no change in uterine glycogen was evinced by the p,p'-isomer, o,p'-DDT increased glycogen content to 385 per cent of the control value.

Welch *et al.*[4] reported increases in the uterine weights of immature rats treated with purified o,p'-DDT as well as technical grade DDT, purified methoxychlor or p,p'-DDT.

TABLE 1. ESTROGEN-LIKE EFFECTS OF p,p'-DDT AND o,p'-DDT ON GLYCOGEN CONTENT AND SEVERAL CARBOHYDRATE-METABOLIZING ENZYMES IN THE UTERUS*

Treatment	Uterine wt. (mg)	Glycogen (μg)†	HK	PFK	Aldolase	PK	G6-PDH	6-PGDH
Control	77 ± 2 (100)	133 ± 27 (100)	3.2 ± 0.2 (100)	7.7 ± 0.3 (100)	9.2 ± 0.4 (100)	190.4 ± 0.3 (100)	6.6 ± 0.1 (100)	2.9 ± 0.1 (100)
p,p'-DDT	88 ± 3 (113)	129 ± 21 (93)	5.3 ± 0.1 (165)‡	9.2 ± 0.3 (119)‡	12.5 ± 0.1 (136)‡	278.6 ± 5.4 (150)‡	9.0 ± 0.1 (136)‡	3.2 ± 0.1 (110)
o,p'-DDT	135 ± 2 (175)‡	513 ± 46 (385)‡	8.6 ± 0.1 (269)‡	18.0 ± 1.5 (233)‡	25.1 ± 2.1 (273)‡	502.8 ± 60.0 (263)‡	18.8 ± 1.3 (285)‡	5.7 ± 0.6 (197)‡

* Each value represents the mean ± S.E. based on 3 determinations of enzyme activities in uteri pooled from 3 rats. Ovariectomized rats were injected with p,p'-DDT (10 mg/100 g) or o,p'-DDT (10 mg/100 g) i.m., and killed after 16 hr. Data are also given in percentages (in parentheses), taking the values of control rats as 100%. For abbreviations, see text.
† Mean ± S.E. represents four to six glycogen assays in each group.
‡ Statistically significant difference when compared with the values of control rats (P < 0.05).

79

Bitman et al.[5] also found that, whereas p,p'-DDT produced minimal effects, o,p'-DDT was capable of inducing significant increases in oviduct weight and glycogen content in two avian species as well as in uterine glycogen of immature rats. The results of the present study showing enhanced uterine weight and glycogen content in o,p'-DDT-treated ovariectomized rats confirm the findings of Bitman et al.[5] and support the conclusion that o,p'-DDT is the more potent isomer for eliciting uterotrophic effects.

It is well established that administration of estrogenic hormones produces marked increases in the activities of a number of enzymes in the uterus.[18–20] Earlier investigations from this laboratory have demonstrated that estrogens also are capable of stimulating the synthesis *de novo* of several uterine glycolytic and pentose phosphate shunt enzymes.[6–13, 16] Barker and Warren[21] have obtained similar evidence for the regulation of G6-PDH activity by estrogens in this tissue. The presently observed increases in the activities of these enzymes as a consequence of o,p'-DDT treatment are reminiscent of the action of estrogenic hormones on the target organ and may constitute a possible basis for further investigations on the estrogenicity of chlorinated hydrocarbon insecticides.

Time-course of o,p'-*DDT-induced changes.* The sequential changes in uterine weight and enzyme activities after a single dose of 10 mg/100 g of o,p'-DDT are shown in Fig. 1. Significant increases in uterine weight were observed as early as 4 hr (130 %) while maximal increases were obtained 16 hr after administration of the insecticide (175 %). Likewise, peak increases in the activities of the four glycolytic enzymes were obtained at 16 hr with hexokinase being elevated to 263 per cent, phosphofructokinase to 233 per cent, aldolase to 273 per cent and pyruvate kinase to 263 per cent of the control values, respectively. The activities of the two shunt enzymes also were increased significantly at 4 hr and rose progressively to 285 and 197 per cent in the case of G6-

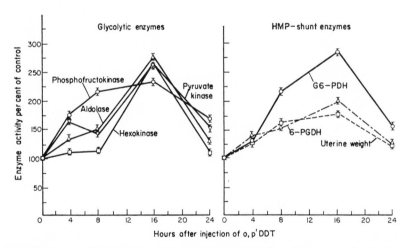

FIG. 1. Time-course of o,p'-DDT-induced increases in uterine weight and enzyme activities. Each point represents the mean and S.E. of three values, each obtained by pooling uteri from three rats. o,p'-DDT (10 mg/100 g) was administered (i.m.) to groups of ovariectomized rats which were killed 4, 8, 16 and 24 hr after injection of the insecticide. Data are given in percentages, taking the values of control rats as 100 %.

PDH and 6-PGDH at 16 hr after o,p'-DDT injection. The present results on o,p'-DDT-induced stimulation of uterine enzymes resemble closely those obtained for the time-course of estradiol induction of uterine hexokinase, phosphofructokinase, aldolase and pyruvate kinase.[6-13] In all cases, statistically significant increases in these uterine enzymes were evident as early as 4 hr after hormone injection, with maximal increases being achieved at 16 hr.

Influence of actinomycin and cycloheximide. The use of inhibitors of RNA and protein synthesis has provided considerable insight into the mechanism of action of a variety of hormones and drugs at the molecular level.[22-24] In order to investigate the nature of the o,p'-DDT-induced enzyme increases, the effects of actinomycin and cycloheximide were examined. The results illustrated in Fig. 2 demonstrate that

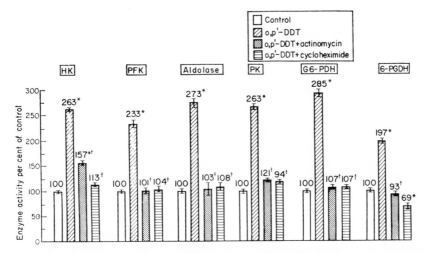

FIG. 2. Effect of actinomycin and cycloheximide on o,p'-DDT-induced increases in uterine enzymes. Bars represent the means and S.E. of three determinations of enzyme activity. Each assay was performed using uteri pooled from three rats. Ovariectomized rats were injected with o,p'-DDT (10 mg/100 g, i.m.) and killed after 16 hr. Actinomycin (25 µg/100 g) or cycloheximide (70 µg/100 g) was administered (i.p.) 30 min prior to o,p'-DDT injection. Data are expressed in percentages, taking the values of control rats as 100%.
* Statistically significant difference when compared with the values of control rats (P < 0·05)
† Statistically significant difference as compared with the values of o,p'-DDT-treated animals without actinomycin or cycloheximide administration (P < 0·05).

actinomycin, which is known to bind to DNA and block DNA-directed synthesis of nuclear RNA,[25, 26] significantly inhibited the insecticide-induced enzyme increases in the uterus. Cycloheximide, which blocks protein synthesis by inhibiting either the transfer of aminoacyl transfer-RNA to ribosomes or the formation of peptide bonds,[27, 28] prevented completely the o,p'-DDT-induced enzyme increases, suggesting that both new RNA and protein synthesis may be involved in the observed o,p'-DDT-stimulation of the uterine enzymes investigated.

Our earlier studies have shown that actinomycin is capable of inhibiting effectively the estrogen-induced increases in hexokinase, phosphofructokinase, aldolase and pyruvate kinase in uteri of ovariectomized animals.[6-12] The ability of actinomycin to prevent the *o,p'*-DDT-induced increases in enzyme activities suggests that, in analogy to estrogenic hormones, the insecticide-stimulated alterations in uterine enzymes may also involve participation of messenger-RNA synthesis at the gene locus.[6, 7, 29]

Effect of cycloheximide during chronic o,p'-*DDT treatment.* The influence of cycloheximide on uterine enzymes in rats treated with *o,p'*-DDT (5 mg/100 g/day) for 3 days is shown in Fig. 3. Chronic treatment with the insecticide resulted in increases

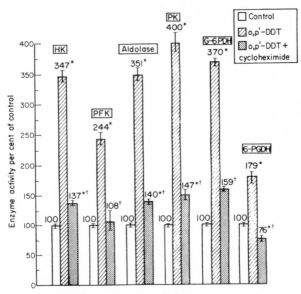

Fig. 3. Influence of cycloheximide administration on uterine enzyme activities in rats treated with *o,p'*-DDT for 3 days. Bars represent the means and S.E. of three determinations of enzyme activity. Each assay was performed using uteri pooled from three rats. Ovariectomized rats were injected with *o,p'*-DDT (5 mg/100 g/day, i.m.) for 3 days, and killed 24 hr after the last injection. Cycloheximide (70 μg/100 g) was administered (i.p.) 30 min prior to the injection of *o,p'*-DDT each day. Data are expressed in percentages, taking the values of control rats as 100%.

* Statistically significant difference when compared with the values of control rats (P < 0·05).
† Statistically significant difference when compared with the values of *o,p'*-DDT-treated animals without cycloheximide administration (P < 0·05).

in the activities of hexokinase, aldolase, pyruvate kinase and G6-PDH that were more pronounced than those observed after a single dose of *o,p'*-DDT (Table 1). Phosphofructokinase and 6-PGDH were increased to virtually the same extent as that seen with a single dose of the pesticide. However, when cycloheximide and the insecticide were injected concurrently for 3 days, the *o,p'*-DDT-stimulated increases in uterine hexokinase, phosphofructokinase, aldolase, pyruvate kinase, G6-PDH and 6-PGDH were significantly inhibited and the values remained close to the normal range. The present results showing greater increases in uterine enzyme activities after chronic *o,p'*-DDT treatment are in accord with our earlier findings demonstrating the greater effectiveness of estradiol in inducing uterine phosphofructokinase and phosphohexose isomerase when administered for 3 days.[6, 7] The inhibition by cycloheximide of the enzyme increases induced by chronic *o,p'*-DDT treatment resembles the inhibitory action of actinomycin on the increases observed in certain uterine enzymes induced by chronic administration of estradiol.[6, 7] Since cycloheximide is a potent inhibitor of protein synthesis, the observed inhibition by this antibiotic of the increases in uterine enzymes elicited by chronic *o,p'*-DDT treatment lends additional support to the view that the augmented uterine enzyme activities involve enzyme synthesis *de novo*.

Influence of progesterone and estradiol. Since *o,p'*-DDT appeared to mimic the effects of estrogens on uterine enzymes, it was of interest to examine the influence of estradiol-17β on the *o,p'*-DDT-stimulated changes in glycolytic and hexose monophosphate shunt enzymes. Administration of estradiol (0·10 μg/100 g) enhanced uterine weights as well as enzyme activities to about the same extent as did the 10·0 mg/100 g dose of *o,p'*-DDT. However, when estradiol and the insecticide were administered concomitantly, the magnitude of uterine enzyme induction, as well as the increases in tissue wet weight, was greater than with either compound injected alone (Table 2). Since progesterone has been shown previously to inhibit the estradiol induction of several uterine enzymes,[6, 7, 10, 12] we considered it pertinent to examine the effects of progesterone on the changes in uterine enzymes induced by *o,p'*-DDT. Administration of progesterone alone resulted in minor but statistically significant increases in uterine wet weight as well as in phosphofructokinase, aldolase, G6-PDH and 6-PGDH. However, when progesterone was administered 30 min prior to the injection of *o,p'*-DDT, the insecticide-induced stimulation of all uterine enzymes was blocked effectively and the values remained close to those observed in control rats. The ability of progesterone to block the estrogen-like effects of *o,p'*-DDT on uterine weight and enzyme activities suggests that this hormone may compete with the insecticide in a manner similar to that reported previously for estradiol-17β.[6, 7, 10, 12]

In contrast to the antagonistic action of progesterone, it is of interest that concomitant administration of estradiol-17β and *o,p'*-DDT produces somewhat additive effects on uterine weights and enzyme activities. Ui and Mueller[30] demonstrated that estradiol stimulates the incorporation of various precursors into uterine RNA and protein. Similarly, Welch *et al.*[4] have shown that treatment of immature female rats with purified *o,p'*-DDT causes a several-fold stimulation of the incorporation *in vitro* of glucose-^{14}C into uterine lipids, protein and RNA. The results of the present study demonstrate that estradiol-17β and *o,p'*-DDT produce additive effects on uterine weights and enzyme activities and indicate further that the hormone and the insecticide educe similar metabolic responses in uterine tissue.

Studies concerning the effects of organochlorine insecticides on reproductive processes in avian and mammalian species have recently been gaining considerable attention. Interference with reproductive performance and a consequent decline in the population of certain avian species have been attributed to environmental pesticide contamination.[1, 3, 31] According to the estimates of Woodwell *et al.*,[32, 33] over one billion pounds (453 $\times$ 10[6] kg) of DDT exists at present in the biosphere. Since technical grade DDT contains *o,p'*-DDT to the extent of 15–20 per cent, almost 200 million pounds of an active estrogen may be present in the environment.[5] Wurster and

TABLE 2. INFLUENCE OF ESTRADIOL-17β AND PROGESTERONE ON o,p'-DDT STIMULATION OF UTERINE ENZYMES*

Treatment	Uterine wt. (mg)	HK	PFK	Aldolase	PK	G6-PDH	6-PGDH
Control	79 ± 2 (100)	3·2 ± 0·2 (100)	7·3 ± 0·2 (100)	11·9 ± 0·2 (100)	165 ± 12 (100)	7·0 ± 0·8 (100)	2·1 ± 0·2 (100)
Estradiol-17β	134 ± 2 (170)†	7·9 ± 0·6 (247)†	21·5 ± 0·6 (294)†	24·2 ± 1·7 (212)†	255 ± 15 (155)†	18·0 ± 0·9 (256)†	4·5 ± 0·3 (214)†
o,p'-DDT	135 ± 2 (171)†	8·6 ± 0·1 (263)†	18·0 ± 1·5 (246)†	25·1 ± 2·1 (212)†	330 ± 15 (200)†	18·8 ± 1·3 (269)†	5·7 ± 0·6 (271)†
Estradiol-17β + o,p'-DDT	165 ± 9 (207)‡	9·7 ± 0·8 (303)‡	29·7 ± 1·2 (407)‡	34·9 ± 1·4 (393)‡	347 ± 15 (210)‡	24·2 ± 2·0 (344)‡	6·0 ± 0·5 (286)‡
Progesterone	98 ± 1 (124)†	3·0 ± 0·1 (94)	10·8 ± 1·4 (148)†	17·7 ± 1·8 (149)†	207 ± 16 (125)	10·7 ± 0·5 (152)†	3·5 ± 0·3 (167)†
Progesterone + o,p'-DDT	80 ± 2 (101)§	2·2 ± 0·3 (70)§	7·9 ± 0·5 (108)§	15·6 ± 1·5 (131)§	192 ± 3 (116)§	9·3 ± 0·1 (132)§	2·8 ± 0·2 (133)§

* Each value represents the mean ± S.E. based on three to four determinations of enzyme activity in uteri pooled from 3 rats. Ovariectomized rats were injected with o,p'-DDT (10 mg/100 g) and sacrificed after 16 hr. Estradiol-17β (0·10 μg/100 g) or progesterone (5 mg/100 g) was also given by the i.m. route 30 min prior to the administration of o,p'-DDT. Data are also given in percentages (in parentheses), taking the values of control rats as 100%. For abbreviations, see text.
† Statistically significant difference when compared with the values of control rats ($P < 0.05$).
‡ Statistically significant difference when compared with the values of rats injected with estradiol alone ($P < 0.05$).
§ Statistically significant difference when compared with the values of rats receiving o,p'-DDT without the administration of progesterone ($P < 0.05$).

84

Wingate[3] have recently suggested that the decline in the population of the Bermuda petrel may be related to the residues of DDT found in the eggs of this species. Ratcliffe[1] related decreased eggshell thickness and weight as well as increased egg breakages to insecticide contamination and suggested that the pesticide may cause a disturbance in the estrogen–parathormone regulation of calcium metabolism in certain birds. In addition, Burlington and Lindeman[2] have demonstrated that DDT produces adverse effects in the cockerel inasmuch as it decreases testicular growth and inhibits the development of secondary sexual characteristics. It has been suggested that o,p'-DDT and the technical grade DDT may exert an estrogenic action on the uterus after undergoing metabolic conversion to an active metabolite by hepatic enzymes. Since most known estrogens possess a phenolic hydroxyl group, it is conceivable that the various analogs of DDT are hydroxylated in the benzene ring to form estrogenic metabolites.[4] Bitman *et al.*[5] and Welch *et al.*[4] have described some estrogen-like properties of o,p'-DDT in the rat, chicken and quail. The results of the present investigation confirm their findings and demonstrate that, like estrogens, o,p'-DDT is capable of augmenting several glycolytic and hexose monophosphate shunt enzymes in the uterus of the ovariectomized rat. These o,p'-DDT-stimulated increases in uterine enzymes can be blocked effectively by actinomycin and cyclo-heximide, two inhibitors of RNA and protein biosynthesis, as well as by progesterone. The presently observed estradiol-like action of o,p'-DDT on several uterine enzymes involved in carbohydrate metabolism suggests the need for closer observation of similar effects that may be exerted by other insecticides.

REFERENCES

1. D. A. RATCLIFFE, *Nature, Lond.* **215**, 208 (1967).
2. H. BURLINGTON and V. F. LINDEMAN, *Proc. Soc. exp. Biol. Med.* **74**, 48 (1950).
3. C. F. WURSTER JR. and D. B. WINGATE, *Science, N.Y.* **150**, 979 (1968).
4. R. M. WELCH, W. LEVIN and A. H. CONNEY, *Toxic. appl. Pharmac.* **14**, 358 (1969).
5. J. BITMAN, H. C. CECIL, S. J. HARRIS and G. F. FRIES, *Science, N.Y.* **162**, 371 (1968).
6. R. L. SINGHAL, J. R. E. VALADARES and G. M. LING, *J. biol. Chem.* **242**, 2593 (1967).
7. R. L. SINGHAL and J. R. E. VALADARES, *Biochem. Pharmac.* **17**, 1251 (1968).
8. R. L. SINGHAL and J. R. E. VALADARES, *Proc. Fourth Int. Congr. Pharmac.* **4**, 196 (1969).
9. R. L. SINGHAL, J. R. E. VALADARES and G. M. LING, *Am. J. Physiol.* **217**, 793 (1969).
10. R. L. SINGHAL and J. R. E. VALADARES, *Am. J. Physiol.* **218**, 321, (1970)
11. J. R. E. VALADARES, R. L. SINGHAL and M. R. PARULEKAR, *Science, N.Y.* **159**, 990 (1968).
12. W. S. SCHWARK, R. L. SINGHAL and G. M. LING, *Biochim. biophys. Acta* **192**, 106 (1969).
13. R. VIJAYVARGIYA, W. S. SCHWARK and R. L. SINGHAL, *Can. J. Biochem.* **47**, 895 (1969).
14. G. E. GLOCK and P. MCLEAN, *Biochem. J.* **55**, 400 (1953).
15. R. L. SINGHAL and G. M. LING, *Can. J. Physiol. Pharmac.* **47**, 233 (1969).
16. R. L. SINGHAL, J. R. E. VALADARES and W. S. SCHWARK, *J. Pharm. Pharmac.* **21**, 194 (1969).
17. S. SEIFTER, S. C. DAYTON, B. NOVIC and E. MUNTWYLER, *Archs. Biochem.* **25**, 191 (1950).
18. A. HERRANEN and G. C. MUELLER, *Biochim. biophys. Acta* **24**, 223 (1957).
19. D. J. MCCORQUODALE and G. C. MUELLER, *J. biol. Chem.* **232**, 31 (1958).
20. B. ECKSTEIN and C. A. VILLEE, *Endocrinology* **78**, 409 (1966).
21. K. L. BARKER and J. C. WARREN, *Endocrinology* **78**, 1205 (1966).
22. H. G. WILLIAMS-ASHMAN, *Cancer Res.* **25**, 1096 (1965).
23. D. W. NEBERT and H. V. GELBOIN, in *Microsomes and Drug Oxidations*, p. 389. Academic Press, New York (1969).
24. R. L. SINGHAL and J. R. E. VALADARES, *Biochem. J.* **110**, 703 (1968).
25. E. REICH, R. M. FRANKLIN, A. J. SHATKIN and E. L. TATUM, *Science, N.Y.* **134**, 556 (1961).
26. T. TAMAOKI and G. C. MUELLER, *Biochem. biophys. Res. Commun.* **9**, 451 (1962).

27. H. L. Ennis and M. Lubin, *Fedn Proc.* **23**, 269 (1964).
28. M. R. Siegel and H. D. Sisler, *Nature, Lond.* **200**, 675 (1964).
29. M. A. Lea, R. L. Singhal and J. R. E. Valadares, *Biochem. Pharmac.*, **19**, 113 (1970)
30. H. Ui and G. C. Mueller, *Proc. natn. Acad. Sci. U.S.A.* **50**, 256 (1963).
31. R. D. Porter and S. N. Wiemeyer, *Science, N.Y.* **165**, 199 (1969).
32. G. M. Woodwell, *Scient. Am.* **216**, 24 (1967).
33. G. M. Woodwell, C. F. Wurster, Jr. and P. A. Isaacson, *Science, N.Y.* **156**, 821 (1967).

The Effect of Environmental and Dietary Stress on the Concentration of 1,1-Bis(4-Chlorophenyl)-2,2,2-Trichloroethane in Rats

J. R. BROWN

The influence of environmental temperature on the action of drugs on isolated tissues is well known. Increased susceptibility of animals to O,O-diethyl O-p-nitrophenyl phosphorothioate (parathion) has been reported by Baetjer and Smith (1956) and by Dehne (1955). Dehne (1955) also has reported on benzene and carbon tetrachloride. Keplinger et al. (1959) studied the effect of ambient temperature on the toxicity of 58 compounds when administered to albino rats. All compounds were most toxic at 36°C with the exception of strychnine, which was equally toxic at 8°C and 36°C, and of promazine and chlorpromazine, which were most toxic at 8°C.

Baetjer et al. (1960) studied the effects of environmental temperature and humidity on lead poisoning in animals. They found that mice exposed to an environmental temperature of 35°C when injected with lead acetate ip or iv began to die sooner, had a higher mortality and shorter average survival time than mice exposed at 22.2°C. Mice with chronic lead poisoning produced by repeated injections of lead acetate or nitrate began to die sooner and had a higher mortality when exposed to a high temperature following cessation of lead injections. Braun and Lusky (1960) reported upon the effect to acclimatization to cold on the action of drugs in the rat. They showed that both pentylenetetrazol and sodium amobarbital were less toxic to rats exposed to 4°C for 24 hr than they were to rats kept at room temperature.

Metcalf (1955) pointed out that 2,2-bis-(p-chlorophenyl)1,1,1-trichloroethane (DDT) was more toxic at low than at high temperatures. Insects treated with the appropriate dose could be cooled to 15°C and thrown into violent convulsions, and when warmed to 35°C appeared entirely normal: this process could be repeated many times. Eaton and Sternburg (1964) showed that the action of DDT on the central nervous system of the *Periplaneta americana* was more toxic at a lower temperature than at a higher temperature, whereas the peripheral nerves showed less change in excitability with change in environmental temperature.

Mobilization of toxic substances stored in the body may occur as a result of illness or metabolic stress. Fitzhugh and Nelson (1947) showed that rats fed high dietary concentrations of DDT (600 ppm or more) developed marked tremors when they were subsequently maintained in a state of starvation. This was thought to be due to the mobilization of DDT stored in the body fat. Dale, *et al.* (1962) studied the storage and excretion of DDT in starved rats. They found that, in rats partially deprived of food after being fed a diet containing 200 ppm DDT, mobilization of body fat resulted in an increased concentration of DDT-derived material in fat tissue and a corresponding increase in the other tissues. An augmented excretion of metabolites occurred during starvation in spite of decreased intake of DDT. The augmented concentration of DDT in the brain, associated with starvation was correlated with the occurrence of signs of poisoning.

The present study concerns the effect of ambient temperature, physical activity, and reduced calorie intake upon the concentration of DDT-derived material in rats. Hayes *et al.* (1958) have shown that in man 2,2-bis-(p-chlorophenyl)-1,1-dichloroethylene (DDE) is the principal storage form of ingested DDT. However, Hayes (1965) has pointed out that rats convert only small amounts of DDT to DDE and monkeys, none at all. Finley and Pillmore (1963) demonstrated the presence of 2,2-bis-(p-chlorophenyl)-1,1-dichloroethane (DDD) in animal tissue obtained from areas where DDT had been sprayed. Datta *et al.* (1964) found that DDD was present in the liver tissues but was absent from the body fat.

It was decided, therefore, to analyze the tissues and blood samples for DDT, DDE, and DDD. The concentration of DDT, DDD, and DDE were determined in the following tissues; blood, brain, muscle, heart, depot fat, liver, kidney, and lung.

METHODS

Male animals of the Wistar strain were used in this study. Their initial weight averaged 250 g. Prior to the administration of DDT the rats were acclimatized to an ambient temperature of either 4°C and 60% relative humidity, or 31°C and 50% relative humidity. During this time they were fed on a synthetic (pesticide free) diet. The composition was as follows: sucrose 64%, casein 17%, corn oil 10%, alphacel 2%, and mineral vitamin supplement 6%. Water was supplied ad libitum.

In all, eight groups of rats were used, with at least four animals to a group. The experimental protocol is given in Table 1.

As far as possible, the rats were maintained in individual cages—this was particularly necessary in the case of rats subjected to cold or to starvation. The animals were placed in water, maintained at the temperatures of the ambient air for each group. They were allowed to swim until they showed signs of being exhausted, when they were removed and dried. The rats were denied food for a period of 1 week. During this time they were given water ad libitum.

At the end of the period of acclimatization the rats were given pure p,p'-DDT dissolved in corn oil, at a dose of 5 mg/kg body weight per day by oral intubation for 14 days. The animals were starved, fed, exercised, or rested according to the experimental design. The period of experimental stress, as outlined above, was continued for a period of 1 week. At the end of this time, the animals were sacrificed. At death a sample of blood was collected from each animal as well as portions of the brain, liver, lung, skeletal muscle, heart, kidney, and depot adipose tissue (subcutaneous). These were immediately placed in separate containers and kept at −79°C until they were analyzed for DDT and its metabolites.

TABLE 1

EXPERIMENTAL PROTOCOL

Experimental conditions	Group	
	4°C	31°C
Feeding and exercise	1	5
Feeding, no exercise	2	6
Starvation and exercise	3	7
Starvation, no exercise	4	8

Analysis of DDT and Its Metabolites

The tissue was weighed and homogenized with chloroform–methanol (2:1) 4 ml/g of tissue for 3 min. The solvent was separated, and the debris was shaken with a further quantity of chloroform–methanol (5 ml/g tissue) for 15 min and filtered. The combined solvent fractions were evaporated under reduced pressure to 2 ml and shaken with acetonitrile saturated with hexane (40 ml). The mixture was diluted with water (250 ml), shaken with hexane (3 × 40 ml), and the combined hexane extracts were dried with anhydrous sodium sulfate.

The hexane solution was evaporated under reduced pressure to 10 ml and chromatographed on a Florisil–Celite column [4:1 (20 g)]. The column was prewashed with hexane, and the pesticides were eluted with 5% ether–hexane (400 ml). The eluate was evaporated to 5 ml under reduced pressure and then to dryness under a gentle stream of nitrogen. The residue was dissolved in hexane (1 ml) and analyzed by gas-liquid chromatography using an electron capture detector.

The identities of DDT and its metabolites were confirmed by thin-layer chromatography. The analytical results have been expressed in terms of μg/100 g tissue.

RESULTS

The concentration of the total DDT-derived material, which includes DDE and DDD is given in Table 2; for rats maintained under different conditions of nutrition, activity and thermal environment.

The concentration of DDT-derived material was greatest in the depot fat and was higher in those animals which were starved than in those which received an adequate caloric intake. No depot fat was found in the exercised and starved rats maintained at an ambient temperature of 4°C. The effect of ambient temperature on the concentration of DDT-derived material in depot fat was equivocal. It was generally higher in those animals maintained at a lower temperature.

TABLE 2

TABLE 2

TOTAL DDT DERIVED MATERIAL (DDT + DDE + DDD) (μg/100 g TISSUE)

| | Starvation | | | | Feeding | | | |
| | Exercise | | No exercise | | Exercise | | No exercise | |
Tissue	4°C	31°C	4°C	31°C	4°C	31°C	4°C	31°C
Blood[a]	38	23	29	19	17	9	11	4
Brain	203	112	206	83	44	42	60	20
Muscle	192	96	155	130	85	77	74	68
Heart	152	168	216	122	52	68	57	34
Depot fat	—	16950	15580	10250	6590	6680	5410	4690
Liver	428	229	341	243	152	101	146	98
Kidney	600	687	477	295	193	87	98	40
Lung	228	230	243	243	224	212	143	69

[a] μg/100 ml blood.

From examination of Table 2 it was found that, in those rats which had been starved, the highest concentrations of DDT-derived material in tissues other than depot fat were in the kidney and the liver. The concentration in the lung was third in order of magnitude, followed by brain, muscle, and heart, in descending order. However, in those animals receiving an adequate dietary caloric intake the highest concentration of DDT-derived material was found in the lung, followed by liver, kidney, muscle, heart, and brain, in descending order of magnitude.

In starved rats, activity and extremes of ambient temperature did not alter the ranking of kidney and liver tissue with respect to organochlorine pesticide concentration. In rats in a state of adequate caloric balance, neither activity nor extremes of ambient temperature appeared to have much effect on the ranking of the lung and liver concentrations of DDT-derived material.

The mean concentrations (μg/100 g) of DDT-derived material for tissues other than depot fat (brain, muscle, heart, liver, kidney, and lung) and for blood, are given in Table 3 for rats maintained under different conditions of caloric balance, activity, and thermal environment.

TABLE 3

TISSUE AND BLOOD CONCENTRATIONS (μg/100 g) OF DDT-DERIVED MATERIAL IN RATS IN RESPONSE TO EXERCISE, STARVATION, AND TEMPERATURE

| | | Starvation | | Feeding | |
Activity	Sample	4°C	31°C	4°C	31°C
Exercise	Tissue[a]	258	217	107	84
	Blood[b]	38	23	17	9
No exercise	Tissue[a]	234	159	84	47
	Blood	29	19	11	4

[a] Brain, muscle, heart, liver, kidney, lung (mean values).

[b] μg/100 ml blood.

It is seen that the concentration of DDT derived material was higher in those groups of rats exposed to the lower ambient temperature, and was higher for the groups under conditions of starvation than for those in a state of caloric balance. The average concentration of DDT derived material in the starved rats was 2.7 times that in the rats in adequate caloric balance when maintained at an ambient temperature of 4°C. When maintained at a temperature of 31°C the ratio was 2.8:1.

TABLE 4

Ratio DDE:DDT Concentration in Tissues and Blood of Rats in Response to Exercise, Starvation, and Temperature

Activity	Sample	Starvation		Feeding	
		4°C	31°C	4°C	31°C
Exercise	Tissue[a]	0.16	0.10	0.13	0.13
	Blood	0.07	0.05	0.16	0.07
No exercise	Tissue[a]	0.18	0.14	0.18	0.16
	Blood	0.13	0.07	0.13	0.09

[a] Brain, muscle, heart, liver, kidney, and lung (mean ratios).

The average ratios of DDE to DDT concentrations in tissues (brain, muscle, heart, liver, kidney, and lung) for rats maintained under different conditions of caloric balance, activity, and thermal environment are given in Table 4 and for DDD:DDT ratios in Table 5. From examination of these tables, it is seen that the ratios of DDD:DDT were somewhat greater for rats exposed at the lower temperature and were lower for rats in a state of caloric balance. The ratios DDD:DDT were generally greater than those for DDE:DDT.

TABLE 5

Ratio DDD:DDT Concentration in Tissues and Blood of Rats in Response to Exercise, Starvation, and Temperature

Activity	Sample	Starvation		Feeding	
		4°C	32°C	4°C	32°C
Exercise	Tissue[a]	0.63	0.36	0.25	0.25
	Blood	0.71	0.16	0.33	0.14
No exercise	Tissue	0.44	0.51	0.43	0.20
	Blood	0.44	0.20	0.25	0.14

[a] Brain, muscle, heart, liver, kidney, and lung (mean ratios).

None of the rats showed any signs of neurotoxicity, such as tremors, convulsions, or coma. None of the animals died as a result of experimental exposure.

The results of the study indicate that there was a marked increase in the tissue concentration of DDT and its metabolites as a result of subjecting rats to starvation.

An increase in ambient temperature has been found to lower the concentration of DDT and its metabolites in the observed tissue. Increased levels of activity were found to result in increased tissue concentrations of DDT and its metabolites.

DISCUSSION

Ecobichon and Saschenbrecker (1967) demonstrated loss of DDT and increase in DDE and DDD from samples of chicken blood as a result of storage at $-20°C$ following a weekly thawing for the removal of subsamples for analysis. Similar findings were presented by Jeffries and Walker (1966) on the breakdown of DDT in avian liver. In the present experiment the samples were stored at $-79°C$, and no degradation of DDT to its metabolites was observed.

After exercise and starvation there was a loss of depot fat, and in general, the results indicated that there was a 3-fold increase in concentration of DDT and its metabolites within the remaining fatty tissue. There may have been, however, a release of these compounds into the general circulation during the exercise period. In the present study environmental and metabolic stress produced an increase in the tissue levels of DDT and its metabolites. However, under the present experimental conditions saturation concentrations in the body fat were not reached and the animals did not exhibit any signs of general toxicity. The liver in the starved rats contained lower concentrations of DDT and its metabolites than did the kidney. However, the DDD content of the liver (242 and 50 $\mu g/100$ g tissue, respectively) in cold acclimatized and starved rats was found to be greater than that in kidney, where the DDT concentration was 84 and 384 $\mu g/100$ g tissue, respectively.

In those rats maintained on an adequate caloric intake, both heat and cold acclimatized, a higher concentration of DDT and its metabolites were found in the lungs than in either the liver or kidneys. This was largely made up of DDT, the concentrations of DDE and DDD were about one-fifth that of DDT. A possible explanation is that lipid soluble materials were carried to the lungs by the chylomicrons, and some of the materials were retained due to the high lipid content of the lungs.

In the rat for tissues other than fat, the DDE concentrations were found to be one-sixth those of DDT, whereas in the fat the concentration was only one-twentieth that of DDT. Brown (1967) found that in man the ratio of DDE:DDT in fat was about 2:1. These observations in general support the evidence of Hayes *et al.* (1958) that insofar as fat is concerned, rats convert only small amounts of DDT to DDE.

The livers of the starved animals contained much less DDT and much more DDD than those of the animals kept on an adequate diet [DDT (starved:fed) 28:68 and DDD (starved:fed) 57:22]. These figures have been corrected for difference in molecular weight. The DDE levels were not significantly different.

Miskus *et al.* (1965) showed that the conversion of DDT to DDD could be accomplished by reduced porphyrins. Our results may reflect a higher percentage of porphyrins in the livers of the starved rats, or more probably, since these livers contained less lipid material, the DDT was more easily accessible than it was in the livers of the nonstarved animals.

The present study was designed to demonstrate possible differences of concentration of DDT and its metabolites as a result of exposure to environmental and metabolic stress. The dose administered was unlikely to cause any toxic reaction under ordinary circumstances and was insufficient to achieve saturation of the depot fat, even under conditions of cold, starvation, and severe physical activity.

REFERENCES

BAETJER, A. M., and SMITH, R. (1956). Effect of environmental temperature on reaction of mice to parathion, an anticholinesterase agent. *Amer. J. Physiol.* **186**, 39–56.

BAETJER, A. M., JOADAR, S. N. D., and McQUARRY (1960). Effect of environmental temperature and humidity on lead poisoning in animals. *Arch. Environ. Health* **1**, 463–477.

BRAUN, H. A., and LUSKY, L. M. (1960). The effect of acclimatization to cold on the action of drugs in the rat. *Toxicol. Appl. Pharmacol.* **2**, 458–463.

BROWN, J. R. (1967). Organo-chlorine pesticide residues in human depot fat. *Can. Med. Ass. J.* **97**, 367–373.

DALE, W. E., GAINES, T. B., and HAYES, W. J. (1962). Storage and excretion of DDT in starved rats. *Toxicol. Appl. Pharmacol.* **4**, 89–106.

DATTA, P. R., LAUG, E. P., and KLEIN, A. K. (1964). Conversion of p,p'-DDT to p,p'-DDD in the liver of the rat. *Science* **145**, 1052–1053.

DEHNE, E. J. (1955). The effects of environmental temperature upon susceptibility to toxic industrial agents. Thesis submitted to School of Hygiene and Public Health, The Johns Hopkins University, 1955.

EATON, J. L., and STERNBURG, J. (1964). Temperature and the action of DDT on the nervous system of Periplaneta americana (L). *J. Inst. Physiol.* **10**, 471–485.

ECOBICHON, D. J., and SASCHENBRECKER, P. W. (1967). Dechlorination of DDT in frozen blood. *Science* **156**, (3775), 663–665.

FINLEY, R. B., and PILLMORE, R. E. (1963). Conversion of DDT to DDD in animal tissue. *Amer. Inst. Biol. Sci.* **13** (3), 41–42.

FITZHUGH, O. G., and NELSON, A. A. (1947). The chronic oral toxicity of DDT [2,2 bis(p-chlorophenyl)-1,1,1-trichloroethane]. *J. Pharmacol. Exp. Ther.* **89**, 18–30.

HAYES, W. J. (1965). Review of the metabolism of chlorinated hydrocarbon insecticides especially in mammals. *Amer. Rev. Pharmacol.* **5**, 27–52.

HAYES, W. J., QUINBY, G. E., WALKER, K. C., ELLIOTT, J. W., and UPHOLT, W. M. (1958). Storage of DDT & DDE in people with different degrees of exposure to DDT. *AMA Arch. Ind. Health* **18**, 398–406.

JEFFRIES, D. L., and WALKER, C. H. (1966). Uptake of pp'DDT and its postmortem breakdown in the avian liver. *Nature (London)* **212**, 533–534.

KEPLINGER, M. L., LANIER, G. E., and DEICHMANN, W. B. (1959). Effects of environmental temperature on the acute toxicity of a number of compounds in rats. *Toxicol. Appl. Pharmacol.* **1**, 156–161.

METCALF, R. L. (1955). Cited by O'Brien, R. D. in: *Insecticides. Action and Metabolism*, p. 332. Academic Press, New York, 1967.

MISKUS, R. P., BLAIR, D. P., and CASIDA, J. E. (1965). Conversion of DDT to DDD by bovine rumen fluid, lake water, and reduced porphyrins. *J. Agr. Food Chem.* **13**, 481–483.

Induction of Enzymes in Mammalian Tissues

DDT-Induced Stimulation of Key Gluconeogenic Enzymes In Rat Kidney Cortex

S. Kacew, R. L. Singhal, and G. M. Ling

Recent investigations have shown that the extensive use of chlorinated hydrocarbon insecticides may be associated with a variety of adverse tissue reactions. In rats, Fahim et al. (1) demonstrated that the depression of the growth rate of neonates of DDT-treated mothers was proportional to the amount of maternally administered insecticide. Damage to liver (2, 3) and heart (4) as well as central nervous system effects characterized by hyperexcitability, tremor, and convulsions (5, 6) have also been reported after the administration of chlorinated hydrocarbon pesticides. Recently, Singhal et al. (7) found that DDT exerted estrogen-like stimulating effects on glycogen and certain key glycolytic and hexose monophosphate shunt enzymes in uteri of ovariectomized rats. Certain chlorinated hydrocarbons also produced renal damage and impaired kidney function as indicated by increased excretion of protein and glucose in the urine (8, 9). Since the observed elevation in urinary glucose excretion after treatment with halogenated hydrocarbons might be related to enhanced renal gluconeogenesis, we examined the influence of DDT on the process of gluconeogenesis in this tissue. Results of our study demonstrate that treatment with DDT increases the activities of renal pyruvate carboxylase, phosphoenolpyruvate carboxykinase, fructose-1,6-diphosphatase, and glucose-6-phosphatase and that the stimulation of these key gluconeogenic enzymes in rat kidney cortex is independent of adrenal function.

Materials and Methods

Experiments were carried out in female Sprague–Dawley rats weighing approximately 100 g and maintained on Master Laboratory Chow and water *ad libitum*. Technical DDT, *o,p'*-DDT, and *p,p'*-DDT were dissolved in corn oil and administered daily by the intramuscular route in a dose of 10.0 mg/100 g. Control animals received an equal volume of corn oil. Some animals were adrenalectomized bilaterally under light pentobarbital anesthesia and used 8 days after the surgery. Adrenalectomized rats received postoperative care as described in an earlier communication (10). All animals were killed by decapitation, kidneys excised, and their cortices carefully dissected and rapidly weighed. The cortical tissue was homogenized in 0.15 M KCl, pH 7.4, and 5% homogenates and supernatant fluids (100 000 × g fraction) were obtained as described previously (11, 12). Pyruvate carboxylase (13), phosphoenolpyruvate carboxykinase (14), and fructose-1,6-diphosphatase (15, 16) activities were assayed in the supernate whereas glucose-6-phosphatase was determined using whole homogenate (15, 16). All enzyme assays were carried out under strictly linear kinetic conditions at 37°. Enzyme activities were calculated as micromoles of substrate metabolized per hour per gram of tissue. In each experiment, the activity of pyruvate carboxylase, phosphoenolpyruvate carboxykinase, and fructose-1,6-diphosphatase is expressed as specific activity per milligram protein in the supernate whereas that of glucose-6-phosphatase is given as specific activity per milligram protein in the homogenate. Protein was determined according to the method of Lowry *et al.* (17). The data were evaluated statistically and the significant differences between the means were calculated as *p* values.

Results

Administration of technical DDT for 3 days resulted in a marked enhancement of the activities of key gluconeogenic enzymes in kidney cortex (Table 1). Following this treatment, pyruvate carboxylase increased to 261%, whereas phosphoenolpyruvate carboxykinase, fructose-1,6-diphosphatase, and glucose-6-phosphatase rose to 200%, 230%, and 314% of the control values, respectively. Since technical DDT contains approximately 20% *o,p'*-DDT and 80% *p,p'*-DDT, the effects of these two isomers on renal gluconeogenic enzymes also

TABLE 1. Influence of DDT administration on key gluconeogenic enzymes in rat kidney cortex

Treatment	PC	PEPCK	FD-Pase	G6-Pase
Control	136.8±11 (100)	14.6±1 (100)	6.2±0.6 (100)	4.1±0.3 (100)
Technical DDT	356.4±48 (261)*	29.1±3 (200)*	14.2±1.5 (230)*	12.9±0.3 (314)*
o,p'-DDT	394.9±31 (289)*	40.1±2.1 (273)*	18.5±1.9 (300)*	12.2±1.9 (298)*
p,p'-DDT	190.0±11 (139)*	29.0±4 (199)*	14.6±0.4 (235)*	7.5±0.5 (183)*

*Statistically significant difference when compared with the values of control rats ($p \leqslant 0.05$).
NOTE: Means ± S.E. represent at least four animals in each group. Rats were treated by injection with technical DDT, o,p'-DDT, or p,p'-DDT, each at a dose of 10.0 mg/100 g, intramuscularly, daily for 3 days and killed 24 h after the last injection. Abbreviations are: PC, pyruvate carboxylase; PEPCK, phosphoenolpyruvate carboxykinase; FD-Pase, fructose-1,6-diphosphatase; G6-Pase, glucose-6-phosphatase. Data are expressed as specific activities per milligram protein and are also given in percentages (in parentheses) taking the values of control animals as 100%.

were investigated. The increases observed following o,p'-DDT administration were comparable to those produced by technical DDT. p,p'-DDT appeared somewhat less effective than the o,p'-isomer, the difference being most marked in the case of pyruvate carboxylase whose activity increased to only 139% of the control values after p,p'-DDT.

Since o,p'-DDT appeared to be more potent in stimulating key gluconeogenic enzymes than the p,p'-isomer, a time-course study was undertaken with o,p'-DDT in order to detect the earliest significant increases in various enzyme activities. Groups of rats were given daily intramuscular injections of 10.0 mg/100 g of o,p'-DDT and killed after 1, 2, 3, 5, and 7 days. Data presented in Fig. 1 show that significant increases in all gluconeogenic enzymes occurred at 2–3 days after the start of o,p'-DDT treatment.

We next investigated the question of whether stimulation of gluconeogenic enzymes observed after DDT was mediated through the release of adrenocortical hormones. Results presented in Table 2 show that the administration of o,p'-

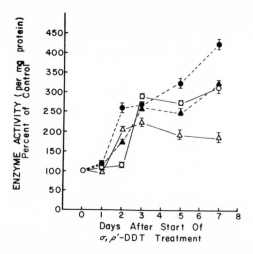

FIG. 1. Time-course of o,p'-DDT-induced stimulation of key gluconeogenic enzymes in kidney cortex. Each point represents mean ± S.E. of at least four rats in each group. Animals were injected with o,p'-DDT (10.0 mg/100 g) intramuscularly, daily for 1–7 days and killed 24 h after the final injection. Enzyme activities are expressed as specific activity per milligram protein and given in percentages taking the value of control animals as 100%. (●) PEPCK, (▲) FD-Pase, (○) PC, (△) G6-Pase.

DDT (10.0 mg/100 g) to adrenalectomized rats produced increases in pyruvate carboxylase, phosphoenolpyruvate carboxykinase, fructose-1,6-diphosphatase, and glucose-6-phosphatase activities which were similar to those observed in the cortices of intact rats. When a potent glucocorticoid, triamcinolone (1.0 mg/100 g), was injected for 3 days, minor but statistically significant increases were seen in the activities of renal pyruvate carboxylase, phosphoenol-pyruvate carboxykinase, and fructose-1,6-di-phosphatase. However, treatment of o,p'-DDT-injected animals with triamcinolone did not augment further the action of o,p'-DDT on the enzymes investigated except in the case of phosphoenolpyruvate carboxykinase. These re-

TABLE 2. Independence of o,p'-DDT-induced stimulation of renal gluconeogenic enzymes from adrenal function

Treatment	PC	PEPCK	FD-Pase	G6-Pase
Control	207.9 ± 12 (100)	18.6 ± 1.9 (100)	7.1 ± 0.5 (100)	6.6 ± 0.2 (100)
o,p'-DDT	490.5 ± 32 (236)*	45.7 ± 2.1 (246)*	18.0 ± 1.2 (255)*	14.7 ± 0.6 (222)*
Triamcinolone	278.3 ± 10 (134)*	27.5 ± 1.1 (148)*	11.6 ± 1.2 (163)*	8.3 ± 0.8 (126)
o,p'-DDT + triamcinolone	473.2 ± 30 (222)*	52.9 ± 0.5 (284)*†	16.9 ± 0.6 (238)*	14.9 ± 1.3 (226)*

*Statistically significant difference when compared with the values of control rats ($p \leqslant 0.05$).
†Statistically significant difference when compared with the values of rats treated with o,p'-DDT alone ($p \leqslant 0.05$).
NOTE: Means ± S.E. represent at least four animals in each group. Adrenalectomized rats were injected with o,p'-DDT (10.0 mg/100 g) intramuscularly, daily for 3 days and killed 24 h after the last injection. Triamcinolone (1.0 mg/100 g for 3 days) was administered either alone or concurrently with o,p'-DDT. Data are expressed as specific activities per milligram protein and are also given in percentages (in parentheses) taking the values of control animals as 100%.

sults suggest that the increases in the activities of key gluconeogenic enzymes produced by o,p'-DDT in the kidney cortex are independent of adrenal function.

Since liver is believed to be the only major tissue besides kidney which is capable of active gluconeogenesis, the effects of o,p'-DDT treatment on the activities of various hepatic key gluconeogenic enzymes also were investigated. A group of rats was given daily intramuscular injections of 10.0 mg/100 g of o,p'-DDT and killed after 7 days. Results presented in Table 3 show that administration of o,p'-DDT resulted in significant increases in all four key gluconeogenic enzymes in hepatic tissue which were generally less pronounced than those noted in the case of kidney cortex enzymes.

Discussion

In mammals, kidney and liver appear to be unique in that they possess the enzymatic potential for glucose synthesis from noncarbohydrate precursors. Pyruvate carboxylase, phosphoenolpyruvate carboxykinase, fructose-

TABLE 3. Influence of o,p'-DDT-treatment on key gluconeogenic enzymes in rat liver

Hepatic enzymes	Control rats	o,p'-DDT-treated rats
PC	531 ± 59 (100)	1030 ± 58 (194)*
PEPCK	7.4 ± 0.6 (100)	24.5 ± 2.6 (331)*
FD-Pase	11.8 ± 1.4 (100)	24.0 ± 0.4 (203)*
G6-Pase	4.1 ± 0.2 (100)	8.6 ± 0.6 (210)*

*Statistically significant difference when compared with the values of control animals ($p \leqslant 0.05$).

NOTE: Means ± S.E. represents four animals in each group. Rats were treated by injection with o,p'-DDT at a dose of 10.0 mg/100 g, intramuscularly, daily for 7 days and killed 24 h after the last injection. Data are expressed as specific activities per milligram protein and are also given in percentages (in parentheses) taking the values of control animals as 100%.

1,6-diphosphatase, and glucose-6-phosphatase are the four enzymes which play a key, rate-limiting role in the reversal of glycolysis since they catalyze irreversible reactions, are involved in circumventing thermodynamic barriers, and are located almost exclusively in organs capable of gluconeogenesis (18). Data presented in this study show that DDT administration increases the activities of all these key gluconeogenic enzymes in the kidney cortex and liver of rats. The observed effects on hepatic and renal gluconeogenesis appeared to be specific since the activities of several key glycolytic and hexose monophosphate shunt enzymes as well as of α-glycerophosphate dehydrogenase and glyceraldehyde phosphate dehydrogenase remained completely unaltered following treatment with this insecticide. Recently, an *acute oral* dose of p,p'-DDT (60.0 mg/100 g), which resulted in marked neurotoxic symptoms, also was found to produce stimulation of pyruvate carboxylase, phosphoenolpyruvate carboxykinase, fructose-1,6-diphosphatase, and glucose-

6-phosphatase activities in rat kidney cortex. Whereas a significant elevation in these enzymes occurred within 1 h, maximal increases were attained 5 h after the insecticide (data to be published). The observed DDT-induced enhancement does not seem to be mediated through the release of adrenocortical steroids since the enzyme increases were not affected by adrenalectomy and concurrent treatment with triamcinolone generally failed to potentiate the action of o,p'-DDT on renal gluconeogenic enzymes.

A close relationship seems to exist between altered renal function and gluconeogenesis in the kidney cortex. Whereas metabolic alkalosis produced by sodium bicarbonate decreased glucose formation, metabolic acidosis observed after the administration of ammonium chloride resulted in a marked enhancement of kidney gluconeogenesis (19, 20). DDT produced tubular degeneration, parenchymous alterations, and vascular congestion in the kidney (3, 21, 22). Klaassen and Plaa (8), using the indicator tape procedure, demonstrated that treatment of mice with certain aliphatic hydrocarbons increased excretion of glucose in the urine. Similarly, in our experiments, daily administration of o,p'-DDT or p,p'-DDT (10 mg/100 g/day) for 7 days was found to elevate the levels of glucose in the urine (more than 2% as indicated on the Lilly Tes-Tape); the increase in urinary glucose (0.25–0.5%) was first detected as early as the 3rd day after insecticide treatment. It is possible that the elevation in urinary glucose levels after DDT might be related to the observed stimulation of renal gluconeogenic enzymes as well as to the tubular damage which results in decreased glucose reabsorption.

1. FAHIM, M. S., BENNETT, R., and HALL, D. G.: Nature Lond. **228**, 1222 (1970).
2. DALE, W. E., GAINES, T. B., and HAYES, W. J., JR.: Toxicol. Appl. Pharmacol. **4**, 89 (1962).
3. DEICHMANN, W. B., KEPLINGER, M., DRESSLER, I., and SALA, F.: Toxicol. Appl. Pharmacol. **14**, 205 (1969).
4. HINSHAW, L. B., SOLOMON, L. A., REINS, D. A., FIORICA, V., and EMERSON, T. E.: J. Pharmacol. Exp. Ther. **153**, 225 (1966).

5. HENDERSON, G. L., and WOOLLEY, D. E.: J. Pharmacol. Exp. Ther. **175**, 60 (1970).
6. HRDINA, P. D., MANECKJEE, A., KACEW, S., PETERS, D. A. V., and SINGHAL, R. L.: Proc. Soc. Can. Fed. Biol. Soc. **14**, 67 (1971).
7. SINGHAL, R. L., VALADARES, J. R. E., and SCHWARK, W. S.: Biochem. Pharmacol. **19**, 2145 (1970).
8. KLAASSEN, C. D., and PLAA, G. L.: Toxicol. Appl. Pharmacol. **9**, 139 (1966).
9. PLAA, G. L., and LARSON, R. E.: Toxicol. Appl. Pharmacol. **7**, 37 (1965).
10. SINGHAL, R. L., and LAFRENIERE, R.: Endocrinology, **87**, 1099 (1970).
11. WEBER, G., SINGHAL, R. L., and STAMM, N. B.: Science, **142**, 390 (1963).
12. WEBER, G., and SINGHAL, R. L.: Biochem. Pharmacol. **13**, 829 (1964).
13. SCRUTTON, M. C., OLMSTED, M. R., and UTTER, M. F.: *In* Methods in enzymology. Vol. 13. *Edited by* S. P. Colowick and N. O. Kaplan. Academic Press, New York, N.Y.
14. PHILLIPS, L. J., and BERRY, L. J.: Am. J. Physiol. **218**, 1440 (1970).
15. SINGHAL, R. L.: J. Gerontol. **22**, 77 (1967).
16. WEBER, G., and SINGHAL, R. L.: J. Biol. Chem. **239**, 521 (1964).
17. LOWRY, O. H., ROSEBROUGH, N. J., FARR, A. L., and RANDALL, R. J.: J. Biol. Chem. **193**, 265 (1951).
18. WEBER, G., SINGHAL, R. L., STAMM, N. B., FISHER, E., and MENTENDIEK, M. A.: Adv. Enzyme Regul. **2**, 1 (1964).
19. GOODMAN, A. D., FUISZ, R. E., and CAHILL, G. F., JR.: J. Clin. Invest. **45**, 612 (1966).
20. ALLEYNE, G. A. O.: Nature Lond. **217**, 847 (1968).
21. SMITH, N. J.: J. Am. Med. Assoc. **136**, 469 (1948).
22. DANOPOULOS, E., MILISSINOS, K., and KATSAS, G.: Arch. Ind. Hyg. Occup. Med. **8**, 582 (1953).

A POSSIBLE ROLE OF LIVER MICROSOMAL ALKALINE RIBONUCLEASE IN THE STIMULATION OF OXIDATIVE DRUG METABOLISM BY PHENOBARBITAL, CHLORDANE AND CHLOROPHENOTHANE (DDT)

M. C. LECHNER and C. R. POUSADA

PRETREATMENT of animals with phenobarbital and many other drugs and foreign substances results in an increased activity of the mixed-function oxidases of the liver, a proliferation of the endoplasmic reticulum and a net increase in its protein content.[1-4]

The exact mechanism of induction remains to be elucidated but evidence based on the use of known inhibitors of protein synthesis and on the incorporation of labeled amino-acids demonstrates that inductive process involves the synthesis of increased amounts of drug metabolizing enzymes. The effects of phenobarbital are prevented by ethionine,[5] puromycin,[6] actinomycin,[7] substances known to block protein synthesis by different mechanisms.

Gelboin and Sokoloff[8] showed that livers from phenobarbital induced animals metabolize drugs at an increased rate and are better able to incorporate amino-acids into microsomal protein *in vitro*. Kato *et al.*[9] showed that the treatment of rats with phenobarbital stimulated the *in vitro* and *in vivo* incorporation of [14C]leucine into microsomal protein. Microsomes from phenobarbital treated rats are more active in L-[14C]phenylalanine incorporation in the absence of polyuridylic acid. After the removal of endogenous messenger RNA by a pre-incubation, the microsomes from phenobarbital treated rats are more than twice as sensitive as control microsomes to

104

polyuridylic acid directed L-[^{14}C]phenylalanine incorporation. Kato et al.[10] have concluded that this can be due to a phenobarbital induced increase in both endogenous microsomal messenger RNA content and the number of ribosomal binding sites available to polyuridylic acid. In fact it is known that phenobarbital treatment affects the total amount of microsomal RNA[7] increasing ribosomal RNA,[11] as an early event produced before any significant increase in microsomal drug metabolizing enzyme activity.[12] These facts have been presented as an explanation for the enhancement in protein synthesis.[13] However, no difference in amino-acid incorporation was observed in ribosomes from control and phenobarbital treated rats indicating that deoxycholate soluble factors, components of the endoplasmic reticulum are important in protein synthesis and are altered by phenobarbital treatment.[10]

Liver microsomes contain an alkaline RNase known to be able to degradate RNA in vitro and presumed to act in vivo.[14] RNase and its cellular inhibitor have been proposed as important control factors in protein synthesis in animal cells in a general way.[15]

In the present work the modifications in alkaline RNase activity in microsomes, isolated from rats during the treatment with phenobarbital and after its suspension were studied. While this investigation was in progress Louis-Ferdinand and Fuller[16] reported a suppression of hepatic RNase in rats during phenobarbital stimulation of drug metabolism.

In order to ascertain if the variations found were a constant factor in the phenomena produced by inducing agents of the phenobarbital group, the alkaline RNase activities were determined in liver microsomes from animals treated with a single dose of chlordane and DDT. Concomitantly the variations in liver weight, aniline hydroxylase, aminopyrine demethylase and cytochrome P-450 were studied in order to establish their time relationships, and possible interdependence.

MATERIAL AND METHODS

Male Wistar rats, 3–3·5 months old were used throughout this investigation. The animals were starved for 24-hr period, receiving water ad lib. before they were weighed and sacrificed. Control and experimental animals were kept under the same experimental conditions, within each experiment. Groups of three rats were used for each sample.

Phenobarbital treated animals were given 80 mg/kg body weight, daily, in aqueous solution (16 mg/ml) intragastrically, for a maximum of 11 days. The rats were killed at different times afterwards (between 12 hr after the first administration and 24 hr after the last, and along 13 subsequent days, after chronic treatment was suspended).

In experiments where other inducing agents have been used, one single dose was given orally, in corn oil solution, 100 mg chlordane, 100 mg DDT/kg body weight and the rats weighed and sacrificed 10, 24 and 48 hr after the administration of each drug.

After decapitation and exsanguination of the animals, livers were quickly removed weighed, pooled and 5 g tissue homogenized immediately in 6·5 vol. of 0·25 M sucrose–1 mM (EDTA). Homogenization and all further operations were carried out at low temperature, and sample containers kept in ice baths.

Liver microsomes were prepared according to the method described by Schenkman et al.[17] by centrifuging the homogenate successively at 500 g for 10 min, 7000 g for

10 min and 12,500 g for 10 min to remove cell debris, mitochondria and light mitochondria. The resultant supernatant was centrifuged for 1 hr at 105,000 g and the microsomal pellet resuspended in an equal volume 0·15 M KCl 25 mM tris buffer (pH 7·5) and resedimented to remove traces of hemoglobin. The washed microsomes were then suspended in 2 vol. of the same buffer mixture, and adjusted to a protein concentration of 6·67 mg/ml determined by the biuret method.

Cytochrome P-450 was determined in microsomal suspensions containing 2 mg protein/ml as described by Remmer et al.[3] from the CO-difference spectrum of dithionite treated microsomes and using the molar extinction coefficient of 91 mM^{-1} cm^{-1} for Δ_A (450–490 mμ) determined by Omura and Sato[18] for calculations and expressed in mμmoles per mg microsomal protein.

Aniline hydroxylase and aminopyrine demethylase activities were determined using an incubation mixture consisting of tris buffer 0·05M (pH 7·5), MgCl$_2$ 5μmoles, Na-isocitrate 8 μmoles, NADP-Na 1 μmole, isocitric deshydrogenase (type IV "Sigma") 20 μl and 8 μmoles of substrate for 1 mg of microsomal protein. Incubations were performed in air, at 37° in a Dubnoff metabolic shaker, for 5 min when the substrate was aminopyrine and for 20 min when aniline was used. Aliquots were precipitated in the cold with TCA, centrifuged and formaldehyde or p-hydroxyaniline formed during the reaction, determined in the supernatant with NASH reagent,[19] (consisting of 150 g ammonium acetate, 3 ml glacial acetic acid, 2 ml acetylacetone, made to 500 ml with distilled water, and pH adjusted to 6·00), and with phenol reagent, respectively.

Aminopyrine demethylase and aniline hydroxylase activities were calculated in mμmoles of HCHO, and in mμmoles p-aminophenol formed per minute per mg microsomal protein.

For the determination of RNase activity, 1·8 ml microsomal suspensions were treated with 0·2 ml of 5% sodium desoxycholate solution, for 30 min at room temperature. An aliquot of 0·1 ml (corresponding to 0·6 mg microsomal protein) was then incubated for 30 min at 37° in a thermostable shaking bath, with 0·25 ml of 0·25 M tris–HCl buffer pH 7·5, 0·1 ml of 20 mM EDTA and 0·25 ml of 1·2% RNA solution. The purified RNA used as substrate was prepared from a commercial RNA (yeast RNA "B.D.H.") that was treated by the method of Kirby[20] and further dialysed against 0·01 M EDTA as described by Shortman.[21]

Immediately after the incubation, the sample tubes were transferred to an ice bath and precipitated with an equal volume of cold 0·75% uranyl-acetate in 25% perchloric acid.

The assays were run in duplicate as well as blank tests. After complete precipitation the soluble fractions were separated by centrifugation at 0° for 30 min at 12,500 g.

0·2 ml of the clear supernatants were diluted with 5 ml of distilled water in a test tube and absorbances determined at 260 mμ. RNase activities were expressed in units per minute per milligram microsomal protein; one unit corresponds to an increase in absolute absorption value of 1000, within the range of linearity.[22]

RESULTS

Rats treated with phenobarbital showed a progressive rise in cytochrome P-450, aminopyrine demethylase and aniline hydroxylase, attaining maximal values at the third day after the initiation of the treatment. Steady state induced conditions were established after the fourth day. Increased values were kept approximatively at the same level until 48 hr after the last phenobarbital administration, decreasing then gradually to normal values.

RNase was significantly reduced 12 hr after the initiation of the treatment reaching lower stable values at 48 hr. After the prolonged phenobarbital treatment was suspended, microsomal alkaline RNase activity showed a sharp rise, as it is represented in Fig. 1.

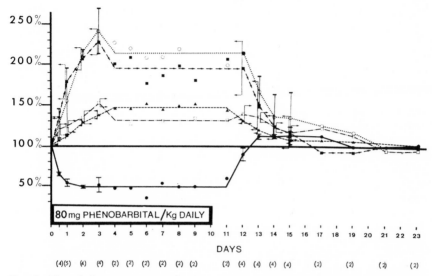

FIG. 1. Plot of the mean percentual values ± S.D. obtained in phenobarbital treated rats, ——●——●——●—— RNase; --▲---▲---▲-- liver weight; →■→→■→→■→ cytochrome P-450 levels; ···○····○····○·· aminopyrine demethylation; -·-□-·-·□-·-·□ -·-aniline hydroxylation. Figures in brackets represent the number of experiments for each point.

Control values were: RNase, 0·177 ± 0·031 units per min per mg microsomal protein; g liver per 100 g body weight, 2·78 ± 0·13; cytochrome P-450, 0·88 ± 0·10 mμmoles/mg microsomal protein; aminopyrine demethylation, 3·14 ± 0·25 mμmoles HCHO liberated per min per mg microsomal protein; aniline hydroxylation 0·178 ± 0·034 mμmoles hydroxy-aniline formed per min per mg microsomal protein.

The yield of microsomal protein recuperated per gram liver was 31·04 mg ± 3·6 for control ($n = 10$). A significant increase was only found in rats treated for 4 consecutive days, when the values reached 130 per cent.

Livers from phenobarbital treated rats are bigger and more friable mainly because of the rise in phospholipids of the membranes and consequent increase in cells size. On account of that, the yield of microsomes and microsomal protein in cell fractionation may be affected, as the behavior of the tissue is different, during homogenization and differential centrifugation. Meanwhile, it can be concluded that the decrease in RNase specific activity observed is not due to a dilution effect as it is produced in the first hours after phenobarbital treatment, when no significant increase in microsomal protein was found (103 per cent for 12 hr treatment, 101 per cent for 24 hr, 104 per cent for 48 hr).

Animals treated with a single dose of chlordane, showed to have significantly increased capacities of oxidative drug metabolism, and P-450 hemoprotein, 24 hr after the administration of the drug, and even more strongly at 48 hr, while RNase activity was reduced to less than 50 per cent of the normal 10 hr after the rats were treated, and during the whole period studied (Fig. 2).

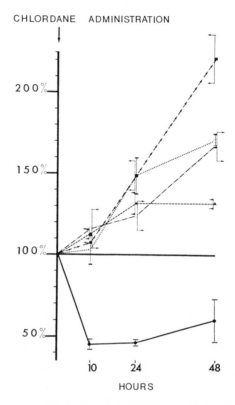

CHLORDANE ADMINISTRATION

HOURS

FIG. 2. Plot of the mean percentual values ± S.D. obtained in chlordane treated rats,——●—●—— RNase; --▲--▲--▲--▲--liver weight; →■→→■→→■→→ cytochrome P-450 levels; ······○······○······○·· aminopyrine demethylation; –·–□–·–·–□–·–□·– aniline hydroxylation.
Control values were: cytochrome P-450, 0·84 ± 0·03 mμmoles; RNase, 0·146 ± 0·023 units; aminopyrine demethylation, 3·27 ± 0·04 mμmoles HCHO; aniline hydroxylation, 0·159 ± 0·015 mμmoles *p*-hydroxyaniline, all expressed per minute per miligram microsomal protein; liver weight 2·40 ± 0·13 g per 100 g body weight.
Each point represents the mean of four experiments.

Ten hours after a single administration of chlorophenothane, a sharp decay in RNase was also produced, and undernormal values maintained along the investigated period. Liver weight, cytochrome P-450 and mixed-function-oxidase activities determined were significantly elevated 24 and 48 hr after the administration of the drug, as it is shown in Fig. 3.

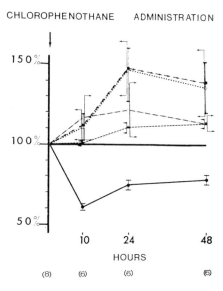

CHLOROPHENOTHANE   ADMINISTRATION

FIG. 3. Plot of the mean percentual values ± S.D. obtained in DDT treated rats,—●—●—●—●— RNase; --▲---▲---▲---▲--- liver weight; → →■→ → →■→ → →■→ →→ cytochrome P-450 levels; ···○···○··· aminopyrine demethylation; —·□·—·—·□— aniline hydroxylation. Control values were: cytochrome P-450 0·87 ± 0·05 mμmoles; RNase, 0·197 ± 0·046 units; aminopyrine demethylation, 3·40 ± 0·49 mμmoles HCHO liberated; aniline hydroxylation, 0·207 ± 0·016 mμmoles p-hydroxyaniline formed, per min per mg microsomal protein. Liver weight 2·61 ± 0·13 g/100 g body weight.
Figures in brackets represent the number of experiments for each point.

As the experiments were performed along a period of several months, the percentage values were determined in relation to controls included in each experiment, in order to prevent the inconvenience of the long-term variations observed in RNase and in oxidative drug metabolism, these recently described by Beuthin and Bousquet.[23]

DISCUSSION

Although the physiological role of RNases is poorly understood, it is known that they are able to hydrolyse cellular RNA's both *in vitro* and *in vivo* mainly the sterically exposed interribosomal segments of messenger RNA, easily available for attack[24] as well as part of the RNA of the larger ribosomal sub-units[25] both involved in binding the ribosomes together in active polysomes.

It is known that in mammalian cells stabilities of polysomes and RNA appear to be at least partly controlled by the activities of RNase[26,27] and it has been shown that RNase is implicated in the regulation of growth. Polysomes prepared from regenerating liver have a lower RNase activity[28] are larger than normal, and more active in supporting protein synthesis *in vitro*.[29] Liver RNase activity also varies with age, being higher in adult than in immature rats according to Arora and de Lamirande,[30] and our own observations (results not included in this publication). Brewer *et al.*[31] have found that the RNase levels of liver postmitochondrial supernatant fractions from hypophysectomized rats were reduced to normal levels by administration of bovine growth hormone when protein-synthesizing capacity of polysomes was restored, and postulated that RNase plays a possible role in the regulation of protein synthesis in rats. Sarkar[14] has found during the action of synthetic corticosteroids an inverse relationship between RNase activity and synthesis of gluconeogenic enzymes.

From the results obtained in the present investigation it can be concluded that:

(1) With all the three agents used, (phenobarbital, chlordane and DDT) the decrease in the specific RNase activity was a constant and early phenomenon, preceding clearly the rise of cytochrome P-450 and activities of the mixed-function-oxidases that were measured.

(2) During phenobarbital chronic administration, when an increased steady state was established, concerning the values of P-450 hemoprotein and related enzyme activities, a constant low level of RNase was maintained.

(3) When prolonged phenobarbital treatment was suspended RNase activity rose sharply reaching values near 100 per cent 48 hr after the last administration, (and rising even slightly until 72 hr), much before normal values of cytochrome P-450, aniline hydroxylase and aminopyrine oxidative demethylase were completely attained.

Our results corroborate those presented by Louis-Ferdinand and Fuller[16] as regards the inhibition of RNase following prolonged administration of phenobarbital. However, while these authors got a complete suppression of RNase activity in rats treated with repeated administration of 100 mg/kg phenobarbital, we have not found such an important reduction with 80 mg/kg even for longer periods. Besides, 3 days after the last administration we have found RNase activities surpassing slightly the normal values, whereas in the mentioned study the activities were still near zero; at this same period aminopyrine demethylase activities were not statistically different from those found in rats killed under treatment while our animals showed a decreased demethylating activity. These divergences may perhaps be explained by the different strain of animals used.

The fact that modifications of RNase activities were inversely related to variations in liver weight, mixed-function-oxidase activities and cytochrome P-450, preceding both, their rise and decay, and reaching stationary levels before induced or normal stable biochemical conditions were established, suggests that RNase may possibly be considered as one of the conditioning factors of the steady states.

The results obtained in the present investigation, although not providing direct evidences, may be taken as additional arguments supporting the idea that RNase plays a role in the regulation of protein synthesis in mammalian cells,[32,33] hypothesis recently reformulated, based on studies of isolated cells and phytohaemagglutinin transformed lymphocytes,[15] and could be responsible for the accumulation of cellular RNA, mainly due to a decreased rate of degradation,[12] and consequently for the enhancement of protein synthesis produced by phenobarbital and other inducing agents of the same group.

Acknowledgements—The authors wish to thank Prof. Dr. H. Remmer for kindly revising the manuscript, and are grateful for the advice and help received from Dr. F. Peres Gomes. We wish to acknowledge the skilful technical assistance of Mrs. Aline Almeida, and are also indebted to Dr. M. C. Duque de Magalhães for the collaboration in the preliminary stage of this work.

REFERENCES

1. J. R. GILLETTE, *Adv. Pharmac.* **4**, 216 (1966).
2. A. H. CONNEY, *Pharmac. Rev.* **19**, 317 (1967).
3. H. REMMER, R. W. ESTABROOK, J. SCHENKMAN and H. GREIM, *Arch. exp. Path. Pharmak.* **259**, 98 (1968).
4. G. J. MANNERING, *Selected Pharmacological Testing Methods* (Ed. M. BURGER) p. 51. Dekker, New York (1968).
5. A. H. CONNEY, C. DAVISON, R. GASTEL and J. J. BURNS, *J. Pharmac. exp. Ther.* **130**, 1 (1960).
6. A. H. CONNEY and A. G. GILMAN, *J. biol. Chem.* **238**, 3682 (1963).
7. S. ORRENIUS, J. L. E. ERICSON and L. ERNSTER, *J. cell. Biol.* **25**, 627 (1965).
8. H. V. GELBOIN and L. SOKOLOFF, *Science* **134**, 611 (1961).
9. R. KATO, L. LOEB and H. V. GELBOIN, *Biochem. Pharmac.* **14**, 1164 (1965).
10. R. KATO, W. R. JONDORF, L. A. LOEB, T. BENT and H. V. GELBOIN, *Molec. Pharmac.* **2**, 171 (1966).
11. J. SEIFERT and H. REMMER, Com. Intern. Symposium on Microsomes and Drug Oxidation—University of Tübingen—Germany, reported by J. B. SCHENKMAN, *Science* **168**, 612 (1970).
12. A. M. COHEN and R. W. RUDDON, *Molec. Pharmac.* **61**, 540 (1970).
13. J. HOLTZMAN and J. R. GILLETTE, *J. biol. Chem.* **243**, 3020 (1968).
14. N. K. SARKAR, *FEBS Letters* **4**, 37 (1969).
15. N. KRAFT and K. SHORTMAN, *Biochim. biophys. Acta* **217**, 164 (1970).
16. R. T. LOUIS-FERDINAND and G. C. FULLER, *Biochem. biophys. Res. Commun.* **38**, 811 (1970).
17. J. B. SCHENKMAN, H. REMMER and R. W. ESTABROOK, *Molec. Pharmac.* **3**, 113 (1967).
18. T. OMURA and R. SATO, *J. biol. Chem.* **239**, 2379 (1964).
19. T. NASH, *Biochem. J.* **55**, 416 (1953).
20. K. S. KIRBY, *Biochem. J.* **64**, 405 (1956).
21. K. SHORTMAN, *Biochim. biophys. Acta* **51**, 37 (1961).
22. T. UCHIDA and F. EGAMI, *Progress in Nucleic acid Research* (Eds. G. L. CANTONI and R. DAVIES) p. 4. Harper, New York (1966).
23. P. K. BEUTHIN and W. F. BOUSQUET, *Biochem. Pharmac.* **19**, 620 (1970).
24. G. BLOBEL and VAN R. POTTER, *Proc. natn. Acad. Sci. U.S.A.* **55**, 1283 (1966).
25. M. L. FENWICK, *Biochem. J.* **107**, 481 (1968).
26. G. R. LAWFORD and H. SCHACHTER, *Can. J. Biochem.* **45**, 144 (1967).
27. J. G. SILER and M. FRIED, *Biochem. J.* **109**, 185 (1968).
28. D. J. S. ARORA and G. DE LAMIRANDE, *Can. J. Biochem.* **45**, 1021 (1967).
29. K. TSUKADA and I. LIEBERMAN, *Biochem. biophys. Res. Commun.* **19**, 702 (1965).
30. D. J. S. ARORA and G. DE LAMIRANDE, *Archs Biochem. Biophys.* **123**, 416 (1968).
31. E. N. BREWER, L. B. FOSTER and B. H. SELLS, *J. biol. Chem.* **244**, 1389 (1969).
32. J. S. J. ROTH, *Biophys. Biochem. Cytol.* **8**, 665 (1960).
33. T. UTSONOMIYA and J. S. ROTH, *J. cell. Biol.* **29**, 395 (1966).

The Effect of Chlorinated Hydrocarbons on Drug Metabolism in Mice

Janis Gabliks, D.D.S., Ph.D. and Ellen Maltby-Askari, B.S.

The introduction of insecticidal and other new chemicals into our environment leads to adaptive changes in many animals. The process of adaptation to a chemical is accompanied by alteration of many biochemical reactions which subsequently may alter certain body responses to other biologically active agents, such as drugs and carcinogens. In some cases, adaptations have been recognized as complicating factors in evaluating the safety of chemicals in our environment. For example, numerous studies have shown that administration of 2,2-bis-(parachlorophenyl)-1,1,1-trichloroethane (DDT) or its structural analog, 2,2-bis-(parachlorophenyl)-1,1-dichloroethane (DDD), to animals affects drug metabolism by altering the activity of hepatic enzymes which also metabolize barbiturates,[1-7] corticosteroids[7-10] and other compounds. [3,5,11-15]

Some of these studies have indicated a marked difference in the response of various animals to barbiturate hypnosis following the administration of DDT and DDD. Increased metabolism of hepatic microsomal enzymes leading to a reduced barbiturate-induced sleep time after repeated administration of DDT and DDD has been reported in the rat by Hart and Fouts,[3-5]; Ghazal et al.,[8]; Straw et al., [6]; and Azarnoff et al.[7] Reduced sleep time has also been reported in rabbits, hamsters, guinea pigs, and chickens by Grady et al.[16]

In contrast to the findings for these species, barbiturate-induced sleep time was prolonged by both DDT and its derivatives in the dog as reported by Nelson and Woodard;[1] Nichols et al.;[2] Azarnoff et al.;[7] and in the mouse by Hart and Fouts;[5] Cram and Fouts;[13] and Grady et al.[16]

112

Species differences also occur with regard to the susceptibility of the adrenals to DDD. Adrenal atrophy has been induced by DDD in the dog but not in the rat[1,2,7] nor in the mouse.[16]

Analysis of these studies suggests that in some cases the differences in animal responses are not definite characteristics of species, but rather vary with the experimental procedure and depend on the general susceptibility of an animal in relation to the duration of exposure to chlorinated hydrocarbons and to the time of barbiturate administration.

Based on our experimental conditions, we observed in mice an initial prolongation of barbiturate hypnosis followed by a reduced sleep time after administration of DDT and DDD. From our studies we postulated a diphasic response of mice to the administration of insecticidal hydrocarbons.

MATERIALS AND METHODS

The initial effects of DDT and o,p'DDD on the duration of pentobarbital hypnosis were determined after a single intraperitoneal injection of these compounds into mice and guinea pigs. Female adult albino mice of the CD-1 strain (from Charles River Breeding Laboratories, Incorporated, North Wilmington, Massachusetts) and female adult guinea pigs (from Elm Hill Farm, Chelmsford, Massachusetts) were fed Purina chow and water ad libitum. DDT (2,2-bis-(parachlorophenyl)-1,1,1-trichloroethane), technical grade, (Nutritional Biochemicals Corporation, Cleveland, Ohio) or o,p'DDD (2,2-(o-chlorophenyl-parachlorophenyl)-1,1-dichloroethane), (Edcan Laboratories, South Norwalk, Connecticut) were suspended in an aqueous 0.1% methylcellulose solution. Groups of ten mice, each weighing about 24 grams, received a single intraperitoneal (IP) injection of 90 mg of DDT or o,p'DDD/kg of body weight at a selected time interval from 4 to 96 hours prior to pentobarbital hypnosis. The lowest effective dose of DDT that affected sleeping time was determined in other groups of ten mice injected IP 18 hours before pentobarbital with varying concentrations of DDT ranging from 0.45 to 90 mg/kg of body weight.

For a comparison of the acute effects of DDT on sleep time in another species, guinea pigs weighing 350 to 400 grams were injected (IP) with DDT at concentrations of 30 and 60 mg/kg of body weight 18 hours prior to the administration of pentobarbital. In all cases, controls were injected (IP) with 1 ml of 0.1% methylcellulose.

The comparative effects on sleep time of repeated treatments with DDT and o,p'DDD were determined in mice treated over a 14-day period with daily injections of 90, 135, and 180 mg of DDT or o,p'DDD/kg of body weight. Pentobarbital was administered 24 hours after the last injection.

113

To measure sleep time, mice were injected IP with 67 mg/kg of sodium pentobarbital — a constituent of Sedasol (Evsco Pharmaceutical Company, Long Island City, New York) which contained per ml of solution: 60 mg of sodium pentobarbital; 20% propylene glycol; 10% alcohol. In order to inject consistent volumes (0.5 ml/mouse) Sedasol was diluted with 0.85% saline solution. Guinea pigs received sodium pentobarbital doses equivalent to 40 mg/kg of body weight.

The duration of hypnosis was measured between the time of pentobarbital injection and the return of the animal's righting reflex. The return of muscle tonicity or muscle hypersensitivity serves as another index of sleep time measurements.

Pentobarbital Metabolism: Adult mice were injected IP with a single dose of 90 mg/kg of DDT or o,p'-DDD 18 hours before the liver was removed. Five livers from each group (after removal of gall bladders) were pooled and homogenized in cold isotonic 1.15% KCl (2.0 ml/gm of liver) with a Sorvall mixer.

Pentobarbital metabolism in vitro was measured by the method described by Kupfer and Peets.[9] The liver homogenate was centrifuged at 0 C for 20 minutes at 9,000 g. The resulting supernatant containing the liver microsomes was used for pentobarbital metabolism in the following way. A mixture containing 1.0 ml of the liver 9,000 g preparation, 1.93 μmol of pentobarbital (Sodium Pentobarbital from Abbott Laboratories, North Chicago, Illinois), 2.0 μmol of NADP, 40.0 μmol of glucose-6-phosphate, 50.0 μmol of nicotinamide, 75.0 μmol of

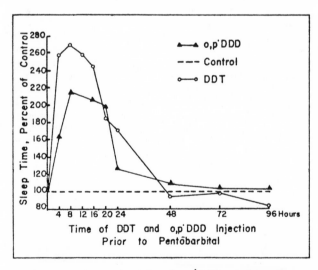

FIGURE 1. Effects of DDT and o,p'-DDD administration on subsequent pentobarbital-induced sleep time in mice. Ten animals per each group were injected (i.p.) with a single dose of 90 mg/kg of either compound at the indicated intervals between 4 and 96 hours prior to hypnosis. Pentobarbital injections were i.p.

114

MgCl$_2$, and a phosphate, buffer (200 μmol) at pH 7.4 (total volume/flask, 3.5 ml) was incubated in a Dubnoff metabolic shaking incubator for 60 minutes at 37 C.

The unmetabolized pentobarbital was measured by the Cooper and Brodie method.[17] Samples of 3 ml were acidified with 1.0 ml of 0.5M citric acid (saturated with NaCl), and the pentobarbital was extracted into heptane containing 1.5% isoamyl alcohol, and then into pH 11.0 phosphate buffer. The optical density was read at 245mμ.

RESULTS

Initial effects of DDT and o,p'DDD on sleep time

Figure 1 demonstrates the effect of a single injection of 90 mg/kg of DDT or o,p'DDD on the duration of pentobarbital hypnosis in mice. Mice treated with either compound at intervals of 4 to 24 hours before pentobarbital treatment showed a markedly prolonged sleep time with the maximal increase after eight hours. In the DDT-treated group the sleep time after four hours was 256% of the control values and after eight hours, 270%. The effect of o,p'DDD was less pronounced and the corresponding values were 162 and 215%, respectively. Furthermore, o,p'DDD showed a slightly shorter duration of the prolonged sleep time effect than DDT, as evidenced by a more rapid disappearance of the effect at the 24-hour test. When DDT or o,p'DDD was given 48, 72 and 96 hours before sodium pentobarbital injection, the sleep time was not significantly different from the control values.

The minimal effective dose of DDT that increased sleep time was measured 18 hours after its administration. The results are summarized in Table 1. A slight increase (144%)

**TABLE 1. The Effect of Varying Doses
of DDT on Pentobarbital
Sleep Time in Mice**

Dose of DDT	Average Sleep Time[a]	% of Control
mg/kg	min ± S.D.	
Control	77 ± 11	100
0.45	88 ± 17[b]	114
4.50	111 ± 17	144
22.50	145 ± 32	188
45.00	170 ± 14	220
90.00	161 ± 32	209

[a]Values represent mean of 10 animals/group

[b]Not statistically significant from control

is already evident with 4.5 mg of DDT/kg. A 10-fold increase of DDT (45 mg/kg) showed the greatest effect with an increase of 220% of the control. A 20-fold increase of the minimal concentration (90 mg/kg) gave no further prolongation of sleep time (209%).

A similar effect of DDT on prolongation of sleep time was also demonstrated in guinea pigs as shown in Table 2. The mean increase for a dose of 30 mg of DDT was 123% of the control; for 60 mg, it was 133%.

The prolongation of pentobarbital sleep time after the administration of DDT or o,p'DDD suggested an inhibitory effect of these compounds on hepatic enzymes

TABLE 2. Sleep Time of Guinea Pigs
Treated with DDT 18 hours
Before Pentobarbital Administration

Dose of DDT	Average Sleep Time[a]	% of Control
mg/kg	min ± S.D.	
Control	290 ± 12	100
30.0	359 ± 35	123
60.0	388 ± 18	133

[a]Values represent mean of 10 animals/group

metabolizing pentobarbital. When pentobarbital was measured in liver enzyme preparations in vitro, its metabolism was markedly reduced in the hydrocarbon-treated mice. The results are shown in Table 3. The DDT-treated livers metabolized approximately one-fifth of the pentobarbital metabolized by the control liver; the o,p''DDD-treated livers metabolized one-half.

Chronic effects of DDT and o,p'DDD on sleep time

Since it is well established that the administration of DDT or o,p'DDD increases enzyme levels for barbiturate metabolism in rats, we expected to reduce the pentobarbital sleep time in mice by daily administration of both compounds during a period of 14 days.

At the doses used (90 to 180 mg/kg) the DDT- and o,p'DDD— treated animals lost weight slightly during the first four days. However, on day 14 their weights were comparable to the weights of the control mice.

116

TABLE 3. Reduced Metabolism of Pentobarbital
in Livers from DDT- and o,p'DDD-treated Mice[a]

Compound	Pentobarbital Metabolized[b]	
	μmols/g liver/hr[c]	% of Control
Control	0.54	100
DDT	0.10	18
DDD	0.22	41

[a]Livers were removed 18 hours after injection of a single dose of 90 mg/kg of DDT or o,p'DDD

[b]Sodium Pentobarbital - Pure powder "Nembutal" (Abbott Laboratories, North Chicago, Illinois. Lot No. 807-1306)

[c]3.86 μmols of pentobarbital was added for each gram of liver homogenate.

Figure 2 shows significant reduction of sleep time with all dose levels: with 90 mg of DDT/kg it was 45% of the control values; with 135 mg, 36%; and with 180 mg, 30%.

In contrast to the effects of DDT, o,p'DDD showed a slight increase (120%) in sleep time at the lowest dose (90 mg/kg). The results with the higher doses (135 mg and 180 mg per kg) of o,p'DDD did not differ significantly from the control values.

DISCUSSION

A markedly prolonged pentobarbital sleep time in mice was evident four hours after intraperitoneal administration of 90 mg/kg of DDT or o,p'DDD. The maximal effect with both compounds occurred after eight hours when the sleep time of DDT-treated mice was 270% of the control value and that of the o,p'DDD-treated mice was 215%. The effect began to diminish after 18 hours, and after 24 hours the sleep time was comparable to the controls. Additional studies showed that 45 mg of DDT gave approximately the same effect and that, even with a small dose of 4.5 mg/kg, slight prolongation of sleep time (144%) was induced.

The prolongation of sleep time after a single dose of DDT or o,p'DDD was associated with the inhibition of liver enzymes metabolizing pentobarbital. This was shown by a reduced metabolism of pentobarbital in vitro using enzyme preparations derived from the livers of the DDT- and o,p'DDD-treated animals.

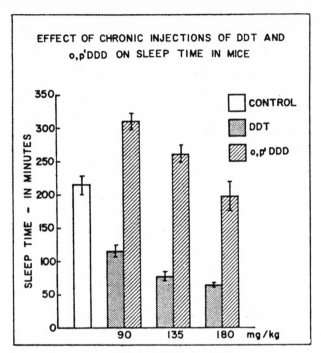

FIGURE 2. Effect of chronic administration of DDT and o,p' DDD on pentobarbital sleep time in mice. Ten animals per each group received 14 daily injections of the compounds at the indicated dose levels. On the day after the last injection of hydrocarbons, pentobarbital (67 mg/kg) was injected (i.p.).

The initial inhibitory effect lasted only up to 24 hours and was dependent on the amount of chlorinated hydrocarbons injected. This interval may coincide with the induction of the processes leading to increased synthesis of the microsomal drug-metabolizing enzymes. There seems to be a certain "latency of onset" in the appearance of enzymes as discussed by Hart and Fouts.[3] This time interval appears to vary in different species. For example, after a single injection of DDT (up to 100 mg/kg) hexobarbital sleep time in the rat is reduced;[5] but not in the mouse when the sleep time is measured on the first, third, or eighth day[5] or on the second day[13] following the treatment. However, data from these reports also indicate a trend toward sleep time prolongation when measured after three hours[3] or when DDT doses above 100 mg/kg were administered.[13] Treatments with 200 or 400 mg/kg of the p,p'DDD derivative given to mice for three days also prolonged hexobarbital sleep time.[16] These observations substantiate our results which show an initial prolongation of pentobarbital sleep time in mice four hours after the administration of DDT or o,p'DDD.

Since a similarity between oxidative drug-metabolizing enzymes and certain steroid hydroxylases has been shown by Kuntzman et al.[18] and by Kupfer and Peets[9] and since Welch et al.[10] demonstrated the inhibitory effect of DDT and other insecticides on the liver microsomal hydroxylation of testosterone in vitro and in vivo, an initial inhibitory effect on steroid metabolism should be considered.

On prolonged administration for 14 days, as shown in this study, the mouse responded to DDT injections with a markedly reduced sleep time, whereas in mice treated with o,p'DDD the sleep time was comparable to the control values. With the lowest dose used (90 mg o,p'DDD/kg) a trend was observed toward a prolongation of this effect. This prolongation of sleep time in mice treated repeatedly with o,p'DDD might be associated with an inhibitory effect on the function of the adrenals. Sleep time can be prolonged by adrenalectomizing rats and conversely curtailed by the administration of glucocorticoids.[20] Adrenal atrophy with o,p'DDD has been shown to occur in dogs[7] and in humans,[19] but has not been established in mice and rats.

A diphasic effect induced by many chemicals leading to an initial inhibition of liver microsomal drug oxidation followed by stimulation of drug metabolism has been reported by Serrone and Fujimoto[21] and Kato et al.[22] Our results suggest that DDT also has a diphasic effect on pentobarbital metabolism, since single doses of both DDT or o,p'DDD markedly prolonged pentobarbital sleep time immediately after administration and DDT also reduced it after repeated injections.

SUMMARY
Chlorinated hydrocarbons, DDT and o,p'DDD, were found to exert a diphasic effect on the hepatic drug metabolizing enzymes in mice and guinea pigs. A single dose of 90 mg/kg of either compound administered within 24 hours of pentobarbital hyponosis resulted in an initial prolongation (270% of the control value) of sleep time. However, repeated injections of the insecticides induced an opposite effect – reduction of sleep time (45% of the control). The effects of DDT on pentobarbital metabolism were dependent on the dose and the length of DDT

119

treatment. The initial prolongation of pentobarbital sleep time by DDT appears to be due to the inhibition of enzymes metabolizing pentobarbital in the liver.

The effects of DDT on immune mechanisms have been investigated in guinea pigs pretreated with DDT and immunized with diphtheria toxoid. Serum antitoxin levels measured by animal protection tests, toxin neutralization in tissue culture, hemagglutination and gel diffusion tests were not significantly different from the control animals. However, the severity of anaphylactic shock, induced by a challenge dose of diphtheria toxoid in the sensitized guinea pigs, was markedly reduced in the DDT treated animals.

ACKNOWLEDGEMENT: We are thankful to Dr. D. Kupfer, Lederle Laboratories, American Cyanamid Company, and to Dr. E. D. Bransome, Massachusetts Institute of Technology, for help and suggestions in the preparation of this manuscript.

REFERENCES

1 Nelson, A. A., and Woodard, G.: Severe adrenal cortical atrophy (cytotoxic) and hepatic damage produced in dogs by feeding 2,2-bis (parachlorophenyl)-l,l, -dichloroethane (DDD or TDE), *Arch. Path.* 48:387-394, 1949

2 Nichols, J., Kaye, S., and Larson, P. S.: Barbiturate potentiating action of DDD and perthane, *Proc. Soc. Exp. Biol. Med.* 98:239-242, 1958

3 Hart, L. G. and Fouts, J. R.: Effects of acute and chronic DDT administration on hepatic microsomal drug metabolism in the rat, *Proc. Soc. Exp. Biol. Med.* 114:388-392, 1963

4 Ghazal, A., et al.: Bescheunigung von Entgiftungsreaktionen Durch Verschiedene Insecticide. *Naunyn-Schmiedeberg Arch. Exp. Path.* 249:1-10, 1964

5 Hart, L. G. and Fouts, J. R.: Further studies on the stimulation of hepatic microsomal drug metabolizing enzymes by DDT and its analogs, *Naunyn-Schmiedeberg Arch. Exp. Path.* 249:486-500, 1965

6 Straw, J.A., Waters, I.W., and Fregly, M.J.: Effect of o,p'DDD on hepatic metabolism of pentobarbital in rats, *Proc. Soc. Exp. Biol. Med.* 118:391-394, 1965

7 Azarnoff, D. L., Grady, H. J., and Svoboda, D. J.: The Effect of DDD on barbiturate and steroid-induced hypnosis in the dog and rat, *Biochem. Pharmacol.* 15:1985-1993, 1966.

8 Kupfer, D., Balazs, T., and Buyske, D. A.: Stimulation by o,p'DDD of cortisol metabolism in the guinea pig, *Life Sci.* 3:959-964, 1964

9 Kupfer, D., and Peets, L.: The effect of o,p'DDD on cortisol and hexobarbital metabolism, *Biochem. Pharmacol.* 15:573-581, 1966

10 Welch, R. M., Levin, W., and Conney, A. H.: Insecticide inhibition and stimulation of steroid hydroxylases in rat liver, *J. Pharmacol. Exp. Ther.*, 155:167-173, 1967

11 Kuntzman, R., Mark L. C., Brand, L., et al.: Metabolism of drugs and carcinogens by human liver enzymes, *J. Pharmacol. Exp. Ther.*, 152:151-156, 1966

12 McLean, A.E.M. and McLean, E. K: The effect of diet and 1,1,1-trichloro-2,2-bis-(p-chlorophenyl) ethane (DDT) on microsomal hydroxylating enzymes and on sensitivity of rats to carbon tetrachloride poisoning. *Biochem. J.* 100:564-570, 1966

13 Cram, R. L. and Fouts, J. R.: The influence of DDT and γ-chlordane on the metabolism of hexobarbital and zoxazoalamine in two mouse strains

Degradative Metabolism of DDT in Mammalian Systems

Degradation of 1,1,1-trichloro-2,2-bis(p-chlorophenyl) ethane by HeLa S cells

E. A. HUANG
J. Y. LU
R. A. CHUNG

DDT* IS AN effective and most widely used insecticide nowadays. The presence of this chlorinated organic compound has been found almost everywhere, including in soils, plants, foods and even in the human body.[1,2]

Persistence of DDT in natural surroundings is well known; however, several *Actinomycetes* were reported to degrade DDT to DDD,[3] and recently Korte et al.[4] also found a number of microorganisms can degrade all representative chlorinated cyclodiene insecticides slowly except dieldrin. The presence of DDT metabolites, DDD and DDE, was found in rat liver,[5] organs of fishes,[6] and in avian liver,[7] indicating that DDT is metabolized in mammalian systems. It is of interest to know how DDT is catabolized in other biological systems. In this experiment, DDT-14C was incubated with HeLa S cells. DDT and its metabolites were extracted with hexane, concentrated, developed by two-dimensional thin-layer chromatography; then the radioactive compounds were identified by autoradiography.

EXPERIMENTAL

HeLa S cells were obtained from the Carver Research Foundation laboratory. Accurately measured cells (4×10^6) were grown in 10 ml of medium (80% medium No. 199, 20% calf serum) which contained 0·05 μc of DDT-14C (specific activity, 2·73 mc/mM, New England Nuclear Corp., Mass.) for periods of 12, 24, 36 and 48 hr at 37°. At the termination of the desired incubation periods, the culture medium was collected by centrifugation. A 10-ml aliquot of *n*-hexane was used to extract DDT and its metabolites from the aqueous phase three times. The combined extract was concentrated to 1 ml by vacuum evaporator. DDT and its metabolites were separated by two-dimensional thin-layer chromatography using Silica gel G-coated plate (2 mm thickness, 20 × 20 cm) as stationary phase. The first developing solvent used was *n*-hexane and a solvent mixture consisting of *n*-heptane, ethanol, acetone at a ratio of 98:0·1:2 was the second-dimensional developing solvent. The resulting chromatogram'was subjected to autoradiography with Kodak, medical X-ray film, NS-54 T for 1 month. The spots appearing on the film were identified by comparing with R_f values of authentic compounds. The portions of silica gel absorbent that correspond to the radioactive spots as judged by the X-ray film were scraped and transferred to a counting vial for radioactivity determination.

Radioactivity was determined in a Packard Tri-Carb liquid scintillation counter. The composition of toluene scintillator fluid was: 4 g BBOT [2,5 bis-(5-*tert*-butylbenzoxazoyl) thiophenol] plus 80 g of napthalene in 400 ml of methyl cellusolve and 600 ml of toluene.

RESULTS AND DISCUSSION

The positions of radioactive spots appearing on the X-ray film are shown in Fig. 1. Five of them were judged as DDE, DBM, DDT, DDD and DBP. Assuming from the low mobility with solvent systems, a spot remaining near the origin could be DDA and BA.

Distribution of radioactivities of DDT and its metabolites at different incubation periods is shown in Table 1. The majority of activities were found in DDT and in DDE. The proportion of radioactivities among DDT, DDE, DBM, DBP, unidentified spot and DDD after 48 hr was 72, 13, 4·9, 4·3, 3·5 and 2·26 respectively. DDT and its metabolites tended to decrease at 12 hr, reached a minimum at 24 hr, then increased again and finally leveled after 36 hr. This trend might reflect growth phases of the cells whereby DDT was absorbed faster together with nutrients in the media by the cell at the initial stage

* Abbreviations: DDT, 1,1,1-trichloro-2,2-bis(p-chlorophenyl) ethane; DDE, 1,1-dichloro-2,2 bis(p-chlorophenyl) ethylene; DDD, 1,1-dichloro-2,2-bis(p-chlorophenyl) ethane; BA, p-chlorobenzoic acid; DDA, bis-(p-chlorophenyl) acetic acid; DDMU, 1-chloro-2,2-bis(p-chlorophenyl) ethylene; DDOH, 2,2-bis-(p-chlorophenyl) ethanol; DBP, 4,4-dichlorobenzophenone; DBM, 4,4-dichlorodiphenyl methane.

124

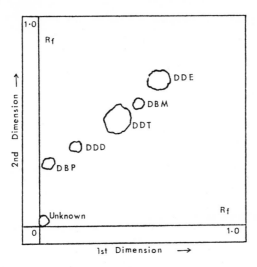

FIG. 1. The degradation pattern of DDT in HeLa S cell system. Solvent No. 1 is *n*-heptane; solvent No. 2, a mixture of *n*-heptane, ethanol and acetone (98:0·1:2 v/v).

TABLE 1. DISTRIBUTION OF DDT AND ITS METABOLITES EXTRACTED FROM HELA S CELL MEDIA AT DIFFERENT INCUBATION PERIODS

	Incubation period (hr)			
Compound	12	24	36	48
Unidentified	82 ± 3*	57 ± 5	80 ± 5	67 ± 2
DBP	76 ± 4	66 ± 1	67 ± 3	82 ± 2
DDD	35 ± 3	19 ± 5	41 ± 1	43 ± 5
DDT	866 ± 8	440 ± 9	1373 ± 2	1364 ± 9
DBM	53 ± 5	10 ± 3	57 ± 2	95 ± 6
DDE	143 ± 3	57 ± 10	212 ± 4	248 ± 2

* Values are expressed as cpm and are mean ± S.E. of six determinations.

of propagation. After reaching a plateau of growth, possibly due to deficiency of some nutrients in the medium, absorption of DDT was decreased and, in the mean time, absorbed DDT and its metabolites were excreted out in the medium from inside the cells.

DDT is a relatively stable compound and its major metabolite is considered to be DDD.[3, 5, 8] It is reported, however, that *Fusarium oxysporum* decomposed DDT to DDD, DDE, DDMU, DDA and DDOH.[9] Guenzi and Beard[10] added DDT-^{14}C to soil and later identified seven radioactive metabolites

In mammals, Vessey *et al.*[8] found DDT, DDD and DDE in cellular fractions of rat liver. Peterson and Robinson[5] reported that rat-liver cells metabolized DDT to DDD and to a lesser extent to DDE. However, they reported that DDE was a terminal metabolite whereas DDD was further metabolized to more polar analogs. Thus once formed, DDE seems to be transferred back to the blood where it is found as the principal metabolite.

In this experiment, five of the radioactive compounds were identified as DDT, DDD, DDE, DBM and DBP. This result shows that many reactions are involved in degradation of DDT in HeLa S cells rather than merely a dechlorination reaction to DDD. The presence of larger amounts of DDE may suggest that DDE is a terminal metabolite of DDT.[5] It is also possible that the presence of iron porphyrin complexes in the medium might have enhanced dehydrochlorination of DDT,[11] resulting in accumulation of DDE.

Acknowledgement—This work was supported by Grant No. RO 1-ES 00095 from the National Institute of Health.

REFERENCES

1. V. FISEROVA-BERGERORA, J. L. RADOMSKI, J. E. DAVIS and J. H. DAVIS, *Ind. Med. Surg.* **36**, 65 (1967).
2. J. GABLIKS and L. FRIEDMAN, *Proc. Soc. exp. Biol. Med.* **120**, 163 (1965).
3. C. L. CHACKO and J. L. LOCKWOOD, *Science, N.Y.* **154**, 893 (1966).
4. F. KORTE, G. LUDWIG and J. VOEL, *Ann. Chem.* **656**, 135 (1962).
5. J. E. PETERSON and W. H. ROBINSON, *Toxic. appl. Pharmac.* **6**, 321 (1964).
6. J. R. DUFFY and D. O'CONNELL, *J. Fish Res. Bd Can.* **25**, 189 (1968).
7. D. J. JEFFRIES and C. H. WALKER, *Nature, Lond.* **212**, 533 (1966).
8. D. A. VESSEY, L. S. MAYNARD, W. H. BRONN and J. W. STULL, *Biochem. Pharmac.* **17**, 171 (1968).
9. R. ENGST and M. KUJAWA, *Nährung* **11**, 751 (1967).
10. W. D. GUENZI and W. E. BEARD, *Science, N.Y.* **156**, 1116 (1967).
11. E. E. FLECK and H. L. HALLER, *J. Am. chem. Soc.* **68**, 142 (1946).

In Vivo Detoxication of p,p'–DDT via p,p'–DDE to p,p' –DDA in Rats

P. R. Datta, Ph.D.

I t is well-known that p,p'-DDT* detoxifies to p,p'-DDE in humans, and stores in the adipose tissue primarily in this form.[1,2] It has been postulated that DDE is an intermediate in the metabolism of DDT to DDA.[3] Findings contradictory to this have been reported, however, which indicate that the conversion of DDE takes place very slowly, if at all, in rats.[4] Because of the preponderance of DDE in human fat, it seemed of importance to attempt to elucidate the detoxication pathway of DDE using high specific activity ^{14}C-p,p'-DDE. To further facilitate this investigation,

---

*Abbreviations: DDT, 1,1,1-trichloro-2,2-bis(p-chlorophenyl)-ethane; DDE, 1,1-dichloro-2,2-bis(p-chlorophenyl)ethylene; DDD, 1,1-dichloro-2,2-bis (p-chlorophenyl)ethane; DDMU, 1-chloro-2,2-bis(p-chlorophenyl)ethylene, from the abbreviation of the generic dichloro diphenyl monochloro unsaturated derivative of DDD; DDMS, 1-chloro-2,2-bis(p-chlorophenyl)-ethane, from the abbreviation of the dichloro diphenyl mono-chloro saturated analogs of DDT and DDD; DDNU, unsym-bis(p-chlorophenyl)ethylene from the dichloro diphenyl non-chlorinated unsaturated analogs of DDE and DDMU; DDOH, 2,2-bis (p-chlorophenyl)ethanol; DDA, bis(p-chlorophenyl) acetic acid.

DDE-pretreated animals were also used — in addition to control ones — since recent reports[5,6] permit one to hypothesize that DDE itself may enhance the enzymes involved in DDE detoxication. The study reported here indicates that DDE is converted to DDA in rats and elucidates the intermediary steps in the detoxication pathway.

EXPERIMENTAL

Four male rats (125-135 grams) of the Osborne-Mendel strain which had been maintained on a synthetic, pesticide-free diet were divided into two groups, control and DDE-pretreated. The DDE-pretreated group was given three daily intraperitoneal injections of crystalline, pure p,p'-DDE (40 mg/kg) to induce DDE-metabolizing enzymes. Both groups were maintained on a synthetic, pesticide-free diet and water, ad libitum.

Forty-eight hours after the last intraperitoneal injection of p,p'-DDE, control and DDE-pretreated rats were intravenously injected with 4 mg/kg of ^{14}C-p,p'-DDE (specific activity 20 mc/mM). ^{14}C-p,p'-DDE was synthesized from ^{14}C-p,p'-DDT by the method of Haller, et al.[7] The purity of this ^{14}C-p,p -DDE was judged to be 99$^+$% by gas-liquid chromatography. Urine and feces were sampled at three-hour intervals and monitored for ^{14}C-p,p'-DDE and its metabolites by liquid scintillation radioassays. Radioactivity was detected in both the urine and feces from DDE-pretreated rats after 12 hours and from control animals after 24 hours. Upon detection of radioactivity, an aliquot of the urine and feces from control and DDE-pretreated rats was extracted with acidic acetone[8] and radioassayed to confirm the presence of acidic ^{14}C-p,p'-DDE-derived materials. Rats were then sacrificed and the liver and kidneys were removed for analysis.

The radioactivity in the liver, kidneys, urine, feces, and carcass of both control and DDE-pretreated rats was quantitatively extracted with ethyl ether. Triplicate 100 μl aliquots of each extract were removed for liquid scintillation radioassay. The distribution of the radioactivity recovered is shown in Table 1.

Twenty ml aliquots of the ether extracts of the liver, kidneys, urine, and feces were evaporated to dryness and

TABLE 1. Distribution of Radioactivity in Tissues and Excreta From Control and p,p'-DDE-Pretreated Rats

Tissues and Excreta	Control Rats Total Radioactivity $(10_6 \, d.p.m.)$*	Percent of Dose	p,p'-DDE-Pretreated Rats Total Radioactivity $(10_6 \, d.p.m.)$**	Percent of Dose
DDE administered	69	100	69	100
Urine	0.07	0.10	0.1	0.17
Feces	11.0	16.0	19.5	28.3
Liver	0.83	1.2	1.3	1.8
Kidneys	0.14	0.20	0.3	0.38
Carcass	55.41	80.3	46.7	67.7
Total Recovery	67.45	97.8	67.9	98.4

*Based on the mean value of triplicate samples from two Control rats sacrificed 24 hours after an intravenous injection of ^{14}C-p,p'-DDE (4 mg/kg; specific activity 20 mc/mM).

**Based on the mean value of triplicate samples from two DDE-Pretreated rats sacrificed 12 hours after an intravenous injection of ^{14}C-p,p'-DDE (4 mg/kg; specific activity 20 mc/mM).

TABLE 2. Comparison of R_f Values of Authentic Compounds with ^{14}C-p,p'-DDE Detoxication Intermediates

Compound	Authentic Compounds (R_f)		^{14}C-p,p'-DDE Detoxication Intermediates (R_f) Urine		Feces		Kidney		Liver	
	I*	II**	I	II	I	II	I	II	I	II
DDE	0.62		—		—		0.59		0.61	
DDMU	0.55		—		—		0.53		0.54	
DDNU	0.69		—		—		0.68		0.68	
DDOH		0.79		0.78		0.79		0.78		0.80
DDA		0.55		0.54		0.56		0.55		—

*Solvent System I: (immobile phase) 8% (v/v) 2-phenoxyethanol in ether (10)
(mobile phase) 2,2,4 trimethylpentane
**Solvent System II: (immobile phase) 7% (v/v) olive oil in acetone (4)
(mobile phase) 2% (v/v) concentrated NH_4OH in absolute ethanol

TABLE 3. Selected Absorption Bands (cm^{-1}) Suitable for Identification of DDE, DDMU, DDNU, DDOH, and DDA

DDE	DDMU	DDNU	DDOH	DDA
967	1575	1610	3226	2762
855	926	985	1053	1675
825	797	840	772	1209
		(absent)		
793				922

subjected to quantitative "clean-up" methods for non-polar[9] and polar (DDOH and DDA)[4] metabolites. Each sample was then paper-chromatographed for separation of ^{14}C-p,p'-DDE and its metabolites. The solvent systems used[4,10] and the relative R_f values of DDE and its metabolites are indicated in Table 2.

DDE, DDOH, and DDA were identified by comparison with authentic reference compounds*, by both paper chromatography (Table 2), and micro-infrared spectroscopy (Table 3.) A tentative structural identification of the two "unknown" intermediates was assigned on the basis of micro-infrared spectroscopy and deductive reasoning based on a knowledge of organic mechanisms. These compounds corresponded to DDMU and DDNU.[4] DDMU was subsequently synthesized from DDD*, and DDNU from DDMS*, by the same method of preparation as ^{14}C-p,p'-DDE from ^{14}C-p,p'-DDT.[7] Gas-liquid chromatograms of these compounds revealed no detectable impurities.

Experimentally, there was found to be a correspondence between the synthetically prepared DDMU and DDNU and the two "unknown" DDE detoxication intermediates, both with respect to characteristic infrared bands (Table 3) and paper chromatographic R_f values (Table 2). On the basis of these two independent means of identification, the two "unknown" detoxication intermediates were positively identified as DDMU and DDNU.

---

*Pure reference standards of DDE, DDD, and DDA were available in this laboratory; DDOH was custom-synthesized by Eastman Organic Chemicals, Inc., Rochester, N.Y., as was DDMS which was further purified in this laboratory by vacuum distillation (175°C, 2mm Hg) and recrystallization from cold petroleum ether.

The relative distribution of radioactivity among ^{14}C-p,p'-DDE and its metabolites in the liver and kidney of DDE-pretreated and control rats is shown in Table 4. These relative ratios were derived by use of the combined techniques of paper chromatography, paper radiochromatogram strip scanning, and liquid scintillation radioassay.

Once the identity of all the detoxication intermediates had been established, their relative positions in the detoxication pathway were studied. Authentic, reference standards of each compound implicated in the detoxication pathway of DDE to DDA (i.e., DDMU, DDNU, DDOH and DDA) were intravenously injected (200 mg/kg) into separate groups of three control rats (125-135 grams) each, which had been maintained on a synthetic, pesticide-free diet. Twenty-four hours after the intravenous injection (the time experimentally found to be required for the detection of ^{14}C-p,p'-DDA in excreta after ^{14}C-p,p'-DDE intravenous injection to control rats), all the rats were sacrificed and their livers and kidneys excised for analysis.

These tissues were extracted with ethyl ether, "cleaned-up,"[4,9] and paper chromatographed[4,10] for separation of metabolites by the same methods described earlier. Metabolites were identified by comparison of their R_f values with those of the authentic reference compounds. Micro-infrared spectroscopic "fingerprinting" of the paper chromatographic spots was used as an additional means of identification. Estimations of the relative ratios of the different metabolites present were based upon densitometer measurements of paper chromatogram strips. The relative distribution of metabolites is shown in Table 5.

RESULTS AND DISCUSSION

The distribution of radioactivity among the excreta, liver, kidneys, and carcass at the time of sacrifice is shown in Table 1 for both control and p,p'-DDE-pretreated rats intravenously injected with ^{14}C-p,p'-DDE. The percentage recovery of radioactivity indicates that essentially all of the radioactivity was recovered. It is to be noted that the amount of radioactivity in the urine from DDE-pretreated rats (0.17% after 12 hours) was considerably greater than

131

TABLE 4. Ratio of Distribution After Intravenous
Injection of ^{14}C-p,p'-DDE

Compound Injected	Tissues Analyzed*	Type of Rat Used	Ratio of Metabolites**				
			DDE	DDMU	DDNU	DDOH	DDA
^{14}C-DDE***	Liver	Control	6	1	2	trace	–
	Liver	DDE-pretreated	4	1	5	1	trace
	Kidneys	Control	7	1	1	2	1
	Kidneys	DDE-pretreated	3	2	2	5	3

*DDA was the predominant metabolite from both the urine (>92%) and feces
(>84%) after injection of 14C-p,p'-DDE; of the other metabolites, only trace
quantities of DDE and DDOH were detectable.
**Average value obtained from 2 rats/type animal/14C-DDE injected.
***4 mg/kg rat of 14C-p,p' DDE (specific activity 20 mc/mM) was injected.

TABLE 5. Ratio of Distribution After Intravenous Injection
of DDE Metabolites

Compound Injected	Tissues Analyzed*	Ratio of Metabolites**				
		DDE	DDMU	DDNU	DDOH	DDA
DDMU	Liver	–	1	7	trace	–
DDNU	Liver	–	–	20	1	–
DDOH	Liver	–	–	–	11	1
DDA	Liver	–	–	–	–	unchanged
DDMU	Kidneys	–	2	2	3	3
DDNU	Kidneys	–	–	1	2	2
DDOH	Kidneys	–	–	–	2	5
DDA	Kidneys	–	–	–	–	unchanged

*DDA was the predominant metabolite recovered from both the urine (>92%)
and feces (>84%) after injection of each compound; of the other metabolites,
only trace quantities of DDE and DDOH were detectable.
**Data represent the average value obtained from 3 rats/detoxication compound
injected.

that found in the urine from control rats (0.10% after 24 hours). The amount of radioactivity in the liver, kidneys, and feces was also significantly greater for DDE-pretreated rats. This indicates not only that DDE-pretreatment did enhance DDE-metabolizing enzymes, but that the detoxication of DDE by control rats is a very slow process. Since the radioactivity was distributed among the liver, kidneys, and excreta of the control rats in readily measurable quantities, however, it may be said that the use of DDE-pretreated rats was not essential to, but facilitated this study.

A comparison of the paper chromatographic R_f values of authentic reference compounds with those of ^{14}C-p,p'-DDE and its detoxication products found in the urine, feces, kidneys, and liver is shown in Table 2. The correspondence in R_f values served as one means of identifying the ^{14}C-p,p'-DDE detoxication compounds.

The infrared spectra of reference standards of the p,p'-isomers of DDE, DDMU, DDNU, DDOH, and DDA were compared with the spectra of ^{14}C-p,p'-DDE and its detoxication compounds. Selected characteristic bands suitable for use in identification are shown in Table 3. This comparison served as an additional means of confirming the identity of ^{14}C-p,p'-DDE and its detoxication compounds, DDMU, DDNU, DDOH, and DDA.

The relative ratios of DDE and its metabolites in the liver and kidneys after intravenous injection of ^{14}C-p,p'-DDE, DDMU, DDNU, DDOH, and DDA are shown in Tables 4 and 5. By interpretation of these data it was concluded that the pathway of detoxication is DDE → DDMU → DDNU → DDOH → DDA. This conclusion was based upon: (a) the profile, and the relative ratios, of the ^{14}C-metabolites found in the liver and kidneys of control and DDE-pretreated rats after injection of ^{14}C-p,p'-DDE (the small quantities of ^{14}C-p,p'-DDE that metabolized within the time period studied indicates the slowness of this detoxication process); (b) the presence of only DDA in the excreta (with the exception of trace quantities of DDOH and DDE), and the fact that injected DDA was recovered unchanged; and, (c) data indicating the favorable conversion of DDMU to DDNU (1:7) in liver, and DDNU to DDOH (1:2) and DDOH to DDA (2:5) in

133

kidneys. (These data also suggest that conversion of DDE → DDMU → DDNU occurs primarily in the liver, and conversion of DDNU → DDOH → DDA primarily in the kidneys. In vitro evidence for this has also been reported.[11])

Tables 4 and 5 show the profile of metabolites found after injection of ^{14}C-p,p′-DDE and DDMU; DDMS was not present as an intermediary compound. Since the conversion of DDT to DDE by a one-step process has previously been established,[12] the complete pathway of detoxication of DDT via DDE to DDA is believed to be as indicated in Figure 1. This pathway, supported by the experimental data in this paper, indicates that DDE detoxication, or enzymatic degradation, to DDNU, a substituted ethylenic compound, occurs by a two-step process; apparently an aliphatic chlorine is replaced by hydrogen in an oxidation-reduction process similar to

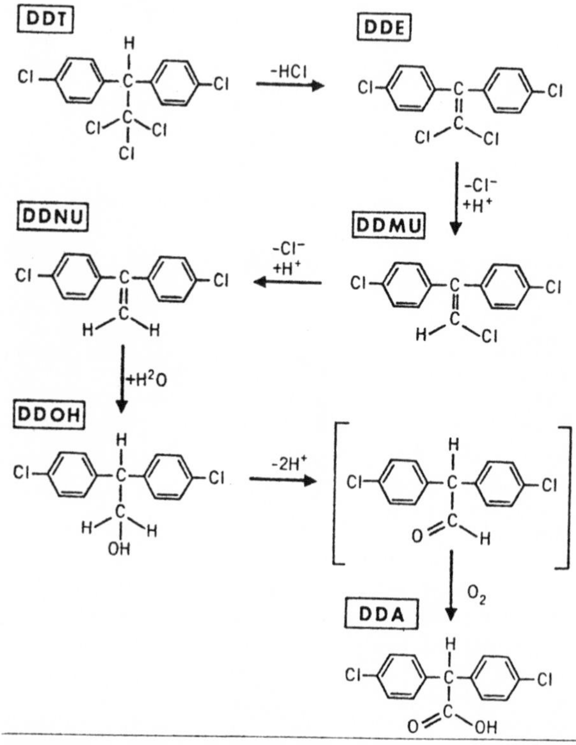

FIGURE 1. Metabolic Sequence for the Conversion of p,p′-DDT via p,p′-DDE to p,p′-DDA in the Rat.

that reported by Castro[13] for the conversion of DDT to DDD by porphyrin complexes. Such a detoxication mechanism may be attributed to hematin-type enzymes.

The schema reported for the conversion of DDT via DDD to DDA in rats[4] indicates that DDMS is an intermediate compound between DDMU and DDNU in that pathway. Since DDMS is absent in the DDT via DDE to DDA pathway reported herein, it appears that DDT, after a one-step conversion to either DDE or DDD' detoxifies by two different pathways in rats. It does not seem unreasonable to speculate that both pathways are operative simultaneously in intact animals; and, that the predominance of either pathway depends upon the physiological response or the amount of toxicant used to challenge the organism.

REFERENCES

1 Mattson AM, et al: Determination of DDT and related substances in human fat *Anal Chem* 25:1065-1070, 1953

2 Hayes WJ Jr, et al: Storage of DDT and DDE in people with different degrees of exposure to DDT, *AMA Archives Indus Health* 18:398-406, 1958

3 White WC, Sweeney TR: A report on the metabolism of DDT, *Public Health Rep*(US) 60:66-71, 1945

4 Peterson JE, Robison WH: Metabolic products of p,p'-DDT in the rat, *Toxicol Appl Pharmacol* 6:321-327, 1964

5 Hart LG, Fouts JR: Further studies on the stimulation of hepatic microsomal drug metabolizing enzymes by DDT and its analogs, *Arch Exp Pathol Pharmakol* 249:486-500, 1965

6 Morello A: Induction of DDT-metabolizing enzymes in microsomes of rat liver after administration of DDT, *Can J Biochem* 43:1289-1293, 1965

7 Haller HL, et al: The chemical composition of technical DDT, *J Amer Chem Soc* 67:1591-1602, 1945

8 Jensen JA, et al: DDT metabolites in feces and bile of rats, *J*

Agric Food Chem 5:912-925, 1957

9 Mills PA, et al: Rapid method for chlorinated pesticide residues in nonfatty foods, *J Assoc Offic Agr Chemists* 46:186-191, 1963

10 Mitchell LC: Separation and identification of chlorinated organic pesticides by paper chromatography, *J Assoc Offic Agr Chemists* 40:294-302, 1957

11 Datta PR, Nelson MJ: p,p'-DDT detoxication by isolated perfused rat liver and kidney, Symposium: Sixth Inter-American Conference on Toxicology and Occupational Medicine, 1968

12 Rothe CF et al: Metabolism of chlorophenothane (DDT), *A M A Archives Indus Health* 16:82-86, 1957

13 Castro CA: The rapid oxidation of iron (II) porphyrins by alkyl halides. A possible mode of intoxication of organisms by alkyl halides, *J Amer Chem Soc* 86:2310-2311, 1964

p,p'-DDT Detoxication by Isolated Perfused Rat Liver and Kidney

P. R. Datta, Ph.D. and M. J. Nelson, Ph.D.

M odification of DDT* in the liver has been inferred from tissue analyses and liver homogenate incubation studies.[1,2] To date, however, no study on the metabolism of DDT by isolated liver and kidney has been reported. Such a study will indicate whether DDT detoxication by the liver and/or kidney is independent of a concomitant or previous effect of DDT on other functioning organs or tissues of the body.

EXPERIMENTAL

Animals

Male weanling rats of the Osborne-Mendel strain were maintained on a synthetic, pesticide-free diet until they reached a size suitable for perfusion studies (250-300 grams).

---

*Abbreviations: DDT, 1,1,1-trichloro-2,2-bis (p-chlorophenyl) ethane; DDE, 1,1-dichloro-2,2-bis (p-chlorophenyl) ethylene; DDD, 1,1-dichloro-2,2-bis (p-chlorophenyl) ethane; DDMU, 1-chloro-2,2-bis (p-chlorophenyl) ethylene, from the abbreviation of the generic dichloro diphenyl monochloro unsaturated derivative of DDD; DDMS, 1-chloro-2,2-bis (p-chlorophenyl) ethane, from the abbreviation of the dichloro diphenyl monochloro saturated analogs of DDT and DDD; DDNU, unsym-bis (p-chlorophenyl) ethylene from the dichloro diphenyl non-chlorinated unsaturated analogs of DDE and DDMU; DDOH, 2,2-bis (p-chlorophenyl) ethanol; DDA, bis (p-chlorophenyl) acetic acid.

Perfusion Apparatus

Both liver and kidney perfusions were carried out in a constant temperature cabinet in a glass apparatus equipped with oxygenator[3] and collecting bulb modifications[4] (Metaloglass Inc., Boston, Mass.). A peristaltic pump with multi-speed transmission (Harvard Apparatus Co., Inc., Dover, Mass.) was used to maintain a constant rate of perfusion through the organ.

Reference Standards

The p,p'-isomers of ^{14}C-DDT, ^{14}C-DDE, DDD, DDMU, DDMS, DDNU, DDOH, and DDA were used for perfusion of isolated rat liver and kidney. ^{14}C-DDT, DDD, and DDA were commercially available in pure form. DDMS and DDOH were custom-synthesized by Eastman Organic Chemicals Inc., Rochester, N. Y.; DDMS was further purified in this laboratory by vacuum distillation (175C, 2mm Hg) and recrystallization from cold petroleum ether. ^{14}C-DDE, DDMU, and DDNU were synthesized[5] from ^{14}C-DDT, DDD, and DDMS, respectively. The non-polar reference compounds (DDT, DDD, DDE, DDMU, DDMS, and DDNU) were judged to be $>99^+\%$ pure by melting point and gas-liquid chromatographic data; the polar reference compounds (DDOH and DDA) were judged to be pure based upon melting point and paper chromatographic data.

ISOLATED LIVER PERFUSION EXPERIMENTS

Hepatectomy

The rats were anesthetized with ether and an abdominal incision was made. The bile duct, portal vein, and inferior vena cava were cannulated in that order. Heparin was used as an anticoagulant. The liver was excised and the tubing connected to the perfusion apparatus. Duration of ischemia (time between occlusion of portal circulation and its re-establishment by perfusion) was 4—6 minutes.

Perfusate

The perfusion fluid used consisted of 67 ml of Weymouth's solution,[6] 33 ml of whole blood obtained from control rats of the Osborne-Mendel strain which had been raised on a synthetic, pesticide-free diet, 2.0 grams of fraction V bovine serum albumin, 825 units of heparin, 6 mg of streptomycin, and 3.5 mg of penicillin G (potassium salt).[7] The total volume of the perfusate was 100 ml.

DDOH and DDA were found to be soluble in the perfusion fluid. Accordingly, 10 mg of DDOH or DDA was dissolved directly in the perfusate prior to perfusion. DDT, DDE, DDD, DDMU, DDMS, and DDNU were found to be only sparingly soluble in the perfusion fluid. Consequently, each of these compounds was made soluble by complexation with egg yolk lipoprotein (mg pesticide/ml lipoprotein stock solution).[8] These complexes were readily soluble in the perfusion fluid and were admixed directly into it. Ten mg of DDD, DDMU, DDMS, or DDNU, or 20 μc of ^{14}C-p,p'-DDT or ^{14}C-p,p'-DDE was used for perfusion.

Perfusion Conditions

The liver was placed in the perfusion apparatus which was set up in a constant temperature cabinet maintained at 37 ± 1C. The perfusion pressure at the hilus of the liver was adjusted to 200 ± 10 mm of perfusate, and this pressure was maintained for the duration of the perfusion. Oxygenation was carried out throughout the perfusion by a mixture of 95% oxygen—5% carbon dioxide under 0.5 atm. of pressure; the gas was humidified by passage through isotonic sodium chloride. The pH of the perfusate was maintained at pH 7.1-7.3 by appropriate addition of 0.5 M sodium bicarbonate ($NaHCO_3$) at hourly intervals, as needed. Each liver was perfused for eight hours.

ISOLATED KIDNEY PERFUSION EXPERIMENTS

Renalectomy

The rats were anesthetized with ether and the peritoneal cavity opened with a long midline incision. The left kidney was freed from its bed, the renal artery exposed and cannulated.[9] Heparin was used as an anticoagulant. The kidney was then removed from the animal and the tubing connected to the perfusion apparatus. Duration of ischemia was 4-6 minutes.

Perfusate

The perfusion fluid was prepared in the same manner as previously described for the liver perfusion experiments. The same quantities of DDT and its metabolites were perfused through the kidney as were perfused through the liver.

139

Perfusion Conditions

A temperature of 37 ± 1C was maintained within the perfusion cabinet for the duration of the perfusion. The pressure through the kidney was adjusted to 80 ± 20 mm of perfusate. Aeration was achieved by 95%:5% oxygen:carbon dioxide under 0.5 atm. of pressure; the gas was humidified by passage through isotonic saline. Each kidney was perfused for a period of six hours.

ANALYSIS OF TISSUES AND PERFUSATES

The liver, kidney, and perfusate from each perfusion experiment were extracted with ethyl ether for the quantitative removal of the compound perfused plus its metabolites. In the case of ^{14}C-p,p'-DDT and ^{14}C-p,p'-DDE, triplicate 100 μl aliquots of each extract were removed for liquid scintiliation radioassay.

A 25-ml aliquot of each ether extract was evaporated to dryness and subjected to quantitative "clean-up" methods for non-polar[10] and polar[2] metabolites. The quantitative recovery of ^{14}C-p,p'-DDT, ^{14}C-p,p'-DDE, and their ^{14}C-metabolites was insured during the extraction of "clean-up" procedures by radioassay of the residue materials in each step of both procedures.

Each sample was paper-chromatographed using solvent systems specific for the separation of non-polar[11] and polar[2] metabolites. Compounds were identified by comparison of their R_f values with those of the authentic reference standards (Table 1). Micro-infrared

TABLE 1. Paper Chromatographic R_f's of DDT and Its Metabolites

Compound	Solvent Systems* 1	2
DDT	0.50	—
DDE	0.62	—
DDD	0.31	—
DDMU	0.55	—
DDMS	0.39	—
DDNU	0.69	—
DDOH	—	0.79
DDA	—	0.55

*1: 8% (v/v) 2-phenoxyethanol in ether   (immobile phase)
    2,2,4 trimethylpentane   (mobile phase)

  2: 7% (v/v) olive oil in acetone   (immobile phase)
    2% (v/v) concentrated $NH_4 OH$ in   (mobile phase)
    absolute ethanol

TABLE 2. Perfusion of DDT and Its Metabolites Through Isolated Rat Liver

Compound Perfused*	Amt. of Perfused compound + Metabolites Rec. from Liver	Distribution of Metabolites in the Liver (%)**							
		DDT	DDE	DDD	DDMU	DDMS	DDNU	DDOH	DDA
14C-DDT (20μc)	0.22 μc	30	10	20	10	5	25	—	—
14C-DDE (20μc)	0.19 μc	—	65	—	10	—	25	—	—
DDD (10mg)	0.15 mg	—	—	40	35	10	15	—	—
DDMU (10mg)	0.12 mg	—	—	—	30	5	65	—	—
DDMS (10mg)	0.10 mg	—	—	—	—	75	25	—	—
DDNU (10mg)	0.09 mg	—	—	—	—	—	100	—	—
DDOH (10mg)	0.13 mg	—	—	—	—	—	—	100	—
DDA (10mg)	0.14 mg	—	—	—	—	—	—	—	100

*The specific activity of the 14C-p,p'DDT and 14C-p,p'DDE used was 4.6 mc/mM and 7.0 mc/mM, respectively.

**Expressed as the average percentage recovered from the liver after eight hours of perfusion; all perfusions were carried out in duplicate. The profile of metabolites in each perfusate showed a similar distribution.

spectroscopic "fingerprinting" of the paper chromatographic spots was used as an additional means of identification.

The total recovery of "compound perfused plus metabolites" from the liver, kidney, and perfusate of each perfusion experiment was determined by liquid scintillation counting for ^{14}C-p,p'-DDT and ^{14}C-p,p'-DDE. For the non-radioactive compounds, standard curves of densitometer optical density readings vs. the number of micrograms spotted were prepared using each of the reference standards. By measuring the optical density of each experimental sample paper chromatographic spot with a densitometer, using the appropriate standard curve to determine the number of micrograms, totaling the number of micrograms recovered for each metabolite, and correcting for dilution factors, it was possible to estimate the total recovery of the non-radioactive "compound perfused plus metabolites" from liver, kidney, and perfusate.

Estimations of the relative quantities of the different metabolities present in liver, kidney, and perfusate were based upon paper radiochromatogram strip scanning and liquid scintillation radioassays for the ^{14}C-compounds, and densitometer measurements of paper chromatogram strips for the non-radioactive compounds. The total recovery and distribution of the "compound perfused plus metabolites" recovered from the liver and kidney of each perfusion experiment are shown in Tables 2 and 3, respectively; although the recovery from the perfusate was much larger, the profile of metabolites showed a similar distribution.

RESULTS AND DISCUSSION

The paper chromatographic R_f values of reference standards of DDT and its detoxication compounds in the solvent systems used for separation[2,11] are shown in Table 1. Metabolites were identified by comparison of their R_f values and micro-infrared spectroscopic "fingerprints" with those of these authentic reference standards.

Data from the isolated perfused liver experiments are shown in Table 2. The total recovery of "compound perfused plus metabolites" from the liver was on the order of 1% in each case. It is interesting to note that

TABLE 3. Perfusion of DDT and Its Metabolites Through Isolated Rat Kidney

Compound Perfused*	Amt. of Perfused compound + Metabolites	Distribution of Metabolites in the Kidney (%)**							
		DDT	DDE	DDD	DDMU	DDMS	DDNU	DDOH	DDA
14C-DDT (20μc)	0.18 μc	100	—	—	—	—	—	—	—
14C-DDE (20μc)	0.21 μc	—	100	—	—	—	—	—	—
DDD (10mg)	0.10 mg	—	—	100	—	—	—	—	—
DDMU (10mg)	0.09 mg	—	—	—	100	—	—	—	—
DDMS (10mg)	0.13 mg	—	—	—	—	30	70	—	—
DDNU (10mg)	0.15 mg	—	—	—	—	—	50	30	20
DDOH (10mg)	0.11 mg	—	—	—	—	—	—	40	60
DDA (10mg)	0.09 mg	—	—	—	—	—	—	—	100

*The specific activity of the 14C-p,p' DDT and 14C-p,p' DDE used 4.6 mc/mM and 7.0 mc/mM, respectively.

**Expressed as the average percentage recovered from the kidney after six hours of perfusion; all perfusions were carried out in duplicate. The profile of metabolites in each perfusate showed a similar distribution.

neither DDNU, DDOH, nor DDA was degraded by the liver. It thus appears that the liver is capable of detoxifying DDT and its metabolites only as far as DDNU. Conversion of DDT, DDD, and DDMU to their degradation products was favorable. However, neither DDE nor DDMS showed a significant amount of conversion. This was not unexpected since it has been reported that the in vivo metabolism of DDE is a slow process,[2,12] and the metabolism of DDMS is more favorable in the kidney than in the liver.[2]

Data from the isolated perfused kidney experiments are shown in Table 3. The total recovery of "compound perfused plus metabolites" from the kidney was also on the order of 1% in each case. It is seen that only DDMS, DDNU, and DDOH are transformed by the kidney to their degradation products. The other perfused compounds were recovered unchanged. In view of the data from the isolated liver perfusion experiments, this would seem to indicate that these three compounds are degraded primarily by the kidney.

The results from the isolated perfused liver and kidney experiments clearly demonstrate that the liver and kidney are two major sites of detoxication of p,p'-DDT. The data indicate that the liver and kidneys each have a specific role in the detoxication of DDT and are capable of carrying out the detoxication processes in an isolated organ. However, the possibility that the liver and kidney of intact animals may further degrade DDT and its metabolites cannot be ruled out.

REFERENCES

1 Datta PR, et al: Conversion of p,p'-DDT to p,p'-DDD in the liver of the rat, *Science* 145:1052-1053, 1964

2 Peterson JE, Robison, WH: Metabolic products of p,p'-DDT in rat, *Toxicol Appl Pharmacol* 6:321-327, 1964

3 Brauer RW, et al: Isolated rat liver preparation. Bile production and other basic properties, *Proc Soc Exper Biol Med* 78:174-181, 1951

4 Flock EV, Owen CA, Jr: Metabolism of thyroid hormones and some derivatives in isolated perfused rat liver, *Am J Physiol* 209:1039-1045, 1965

5 Haller HL, et al: The chemical composition of technical DDT, *J Amer Chem Soc* 67:1591-1602, 1945

6 Weymouth C: Rapid proliferation of sublines of NCTC clone 929 (Strain L) mouse cells in a simple chemically defined medium (MB 752/1), *J Natl Cancer Inst* 22:1003-1017, 1959

7 Juchau MR, et al: The induction of benzpyrene hydroxylase in the isolated perfused rat liver, *Biochem Pharmacol* 14:473-482, 1965

8 Lipke H, Kearns CW: DDT dehydrochlorinase. I. Isolation, chemical properties, and spectrophotometric assay, *J Biol Chem* 274:2123-2128, 1959

9 Skeggs LT, Jr, et al: The preparation and function of the hypertensin-converting enzyme, *J Exp Med* 103:295-299, 1956

10 Mills PA, et al: Rapid method for chlorinated pesticide residues in nonfatty foods, *J Assoc Offic Agr Chemists* 46:186-191, 1963

11 Mitchell LC: Separation and identification of chlorinated organic pesticides by paper chromatography, *J Assoc Offic Agr Chemists* 40:294-302, 1957

12 Datta PR: In vivo detoxication of p,p'-DDT via p,p'-DDE to p,p'-DDA in rats: Symposium, Sixth Inter-American Conference on Toxicology and Occupational Medicine, 1968

Nonconversion of o,p'-DDT to p,p'-DDT in Rats, Sheep, Chickens, and Quail

JOEL BITMAN

HELENE C. CECIL, GEORGE F. FRIES

Klein *et al.* (*1, 2*) reported the supposed isomeric conversion of *o,p'*-DDT to *p,p'*-DDT in the rat (*3*). This conversion was based on the finding of appreciable quantities of *p,p'*-DDT after feeding *o,p'*-DDT to rats. Such a conversion would involve either (*1*) splitting off the *o*-chlorophenyl group from the ethane chain with subsequent recombination to form a *p,p'* molecule or (*2*) replacement of the *o*-Cl by H, and chlorination of the para position. Both of these mechanisms appear to be unlikely biological metabolic reactions (*4*). The purpose of our study was to demonstrate that the conversion of *o,p'*-DDT to *p,p'*-DDT does not occur biologically and to provide a more logical explanation for the appearance of *p,p'*-DDT after feeding pure *o,p'*-DDT.

An explanation for the supposed conversion can be given by the presence of *p,p'*-DDT as an impurity in the *o,p'*-DDT which was fed. When a dilute (1 μg/ml) solution of *o,p'*-DDT (over 99 percent pure, Aldrich) was analyzed by gas-liquid chromatography (GLC) (*5*), a single peak with a retention time characteristic of *o,p'*-DDT is observed. When, however, a concentrated (100 μg/ml) solution was analyzed, the presence of *p,p'*-DDT was also noted. Thin-layer chromatography (TLC) of an equivalent sample of this solution and subsequent GLC analysis resulted in essentially quantitative recovery of *p,p'*-DDT from this supposedly pure *o,p'*-DDT solution. Thus, the simple step of analyzing a concentrated solution of *o,p'*-DDT, rather than an extremely dilute one, demonstrated the presence of *p,p'*-DDT.

During the past 2 years, we have

Table 1. *p,p'*-DDT in rats, sheep, Japanese quail, and chickens after feeding impure *o,p'*-DDT.

Species	*o,p'*-DDT in diet (ppm)	Time (days)	*p,p*-DDT impurity (%)	*p,p'*-DDT intake (mg)	*p,p'*-DDT retained (mg)	Retention (%)
Rat	20	98	0.4	0.118	0.065	55
	40	98	0.4	0.235	0.116	49
	100	22	1.3	0.453	0.322	71
Sheep-ewe	10	120	0.5	8.400	0.625	7
Lamb	10	87	0.5	2.958	1.012	34
Japanese quail	100	45	0.4	0.144	0.104	72
Chicken	150	98	0.6	9.700	5.004	52

conducted experiments in which *o,p'*-DDT was fed to rats (*Rattus norvegicus*), sheep (*Ovis aries* L.), Japanese quail (*Coturnix coturnix japonica*), and chickens (*Gallus domesticus*). Analysis of body lipid revealed that *o,p'*-DDT was the major pesticidal residue but that significant quantities of *p,p'*-DDT were found (Table 1). In all cases, enough *p,p'*-DDT was ingested as an impurity to account for the *p,p'*-DDT found in the animals at the end of the experiment.

Analysis of three batches of commercial *o,p'*-DDT are shown in Table 2. These batches (Aldrich) contained much more *p,p'*-DDT than did earlier samples from the same source, and they contained significantly more *p,p'*-DDT than a sample obtained from the Pesticide Chemicals Branch of the U.S. Department of Agriculture (USDA-ENT 3983). The USDA *o,p'*-DDT has been prepared by isolation from technical DDT 25 years ago in a study of the composition of technical DDT (6).

As additional proof that *p,p'*-DDT is not formed from *o,p'*-DDT, pure *o,p'*-DDT was prepared for subsequent feeding trials. Samples (60 mg) of commercial *o,p'*-DDT (Aldrich) were chromatographed on 150 g of aluminum oxide (Merck) in a glass column [27 mm (inside diameter) by 300 mm]. The chromatogram was developed with 300 ml of *n*-hexane, and each 10-ml fraction was analyzed by GLC for the presence of *o,p'*-DDT, *p,p'*-DDT, and *p,p'*-DDE. Those fractions containing only *o,p'*-DDT, usually fractions 13 to 15, were combined to provide pure *o,p'*-DDT for the feeding trials. By this technique, a product was obtained which was 99.974 percent pure *o,p'*-DDT (USDA-DCRB-1; see Table 2).

Female rats (215 to 230 g) were fed a control diet or a diet containing

Table 2. Analyses of *p,p'*-DDT contamination in samples of *o,p'*-DDT.

Sample	*o,p'*-DDT (%)	*p,p'*-DDE (%)	*p,p'*-DDT (%)
Aldrich			
092381	98.784	0.026	1.190
101591	98.639	0.023	1.338
110407	98.915	0.026	1.059
USDA-ENT			
3983	99.477	0.180	0.344
USDA-DCRB			
1	99.974	0.009	0.017

p,p'-DDT (Aldrich), 100 ppm; *o,p'*-DDT (98.8 percent, Aldrich), 100 ppm; or *o,p'*-DDT (99.974 percent, USDA-DCRB), 100 ppm. After 22 days on the experimental diets, the rats were killed and samples of body lipid and the whole body carcass were analyzed for pesticide residues. *o,p'*-DDT uniformly labeled with ^{14}C (impure mixture, Nuclear-Chicago) was purified by TLC on alumina, and the isolated radioactive *o,p'*-DDT was dissolved in olive oil. Forty-eight hours before the scheduled killing time, the rats receiving the diet containing the Aldrich *o,p'*-DDT received about 5 μc of the ^{14}C-labeled *o,p'*-DDT by stomach tube.

When impure *o,p'*-DDT (98.8 percent, Aldrich) was fed to adult female rats, a sizable amount of *p,p'*-DDT was found in the body lipid (Table 3). The presence of this *p,p'*-DDT could be the basis for a faulty conclusion that this *p,p'*-DDT was formed from the *o,p'*-DDT which was fed. The group fed our purified *o,p'*-DDT (99.974 percent pure) demonstrates conclusively that no conversion occurs (Table 3). The *p,p'*-DDT level in the body fat of these rats was very similar to control levels. If impure *o,p'*-DDT is fed to rats, pesticide residues contain *p,p'*-DDT. If pure *o,p'*-DDT is fed to rats, no *p,p'*-DDT is found.

The lipid extract from the rats fed the impure *o,p'*-DDT and injected with radioactive o,p'-DDT was subjected to TLC on alumina, with hexane being used to develop the chromatogram. In this system, o,p'-DDT has a greater R_F than p,p'-DDT. Approximately 1 percent of the radioactivity recovered from the body lipid was p,p'-DDT (Table 4, column 6), a value very similar to the amount originally present in the radioactive o,p'-DDT which was fed (column 3). In order to demonstrate further the identity of the radioactive TLC spots as o,p'-DDT and p,p'-DDT, the extracts and dosing solution were subjected to dehydrohalogenation in KOH to convert the compounds to o,p'-DDE and p,p'-DDE. The DDE products were then subjected to TLC on alumina in a mixture of 2 percent acetone and hexane. Under these conditions, there is a reversal in the retention times of the o,p'- and p,p'- isomers, and p,p'-DDE has a greater R_F value than o,p'-DDE has. Again, the radioactivity recovered as p,p'-DDE was similar to that found in the dosing solution and provided no evidence of a conversion of radioactive o,p'-DDT to radioactive p,p'-DDT. It is also probable that part of the 1 percent of the radioactivity appearing as p,p'-DDT or p,p'-DDE during the TLC-reversal steps is due to fast-running or tailing of the o,p'-DDT. The 1 percent of the radioactivity in the p,p'-DDT spot is in marked contrast to the amount of unlabeled p,p'-DDT recovered in the body lipid of these same rats which had also been fed the Aldrich o,p'-DDT [28.5 percent of the total DDT residue was present as p,p'-DDT by GLC (Table 3)]. The administration of radioactive o,p'-DDT thus demonstrated, by an additional independent means, that no isomeric conversion of o,p'-DDT to p,p'-DDT occurred.

Since the original reports of Klein et al. (1, 2), there have been two reports which have directly claimed that o,p'-DDT is transformed into p,p'-DDT (7, 8). Ecobichon and Saschenbrecker in 1968 (7) stated that they have shown that o,p'-DDT is converted to p,p'-DDT but no data were given. In 1969 French and Jefferies (8) fed 250 mg of o,p'-DDT to homing pigeons (Columbia livia) and found p,p'-DDT and p,p'-DDE in the body fat. A contamination of 0.5 percent p,p'-DDT in

Table 3. Pesticide residues after feeding p,p'-DDT, impure o,p'-DDT, or pure o,p'-DDT.

Group	Body fat concentrations (μg/g)				
	p,p'-DDE	o,p'-DDD	o,p'-DDT	p,p'-DDD	p,p'-DDT
Control	0.8		0.1		0.6
p,p'-DDT	36.4			5.7	413.8
o,p'-DDT, Aldrich	2.0	2.6	36.6		14.6
o,p'-DDT, purified	0.9	3.5	72.7		1.0

Table 4. Administration of radioactive o,p'-DDT to rats. Values are means of four rats.

Thin-layer chromatography of [^{14}C]DDT	Total dose			Retained in body		
	o,p'-DDT (count/min)	p,p'-DDT (count/min)	p,p'-DDT (%)	o,p'-DDT (count/min)	p,p'-DDT (count/min)	p,p'-DDT (%)
Untreated	10,086,000	135,000	1.3	793,456	6980	0.9
After dehydrohalogenation	10,117,000	93,500	0.9	831,508	5248	0.6

the *o,p'*-DDT fed would have supplied each pigeon with 1250 μg of *p,p'*-DDT, an amount much greater than the total content of *p,p'*-DDT and *p,p'*-DDE recovered.

Two earlier studies failed to yield evidence of the conversion of *o,p'*-DDT to *p,p'*-DDT (*9, 10*). Mendel *et al.* (*9*) attempted to determine the location of the site of the unusual isomeric conversion of *o,p'*-DDT to *p,p'*-DDT. They incubated *o,p'*-DDT both aerobically and anaerobically with *Aerobacter aerogenes*, but recovered only unchanged *o,p'*-DDT and *o,p'*-DDD. They concluded that this coliform organism was not the mediator of the conversion in the gut of the rat. Recently, Lamont *et al.* (*10*) fed mallards (*Anas platyrhynchus*) *o,p'*-DDT, but since there was no increase in *p,p'*-DDT or its metabolites, they concluded that "the biological isomeric transformation evidently did not take place."

We now have demonstrated conclusively that when pure *o,p'*-DDT is fed to rats, there is no conversion to *p,p'*-DDT. In addition, radioactive *o,p'*-DDT fed to rats did not give rise to radioactive *p,p'*-DDT. Conversely, we have also shown that, if impure *o,p'*-DDT containing small amounts of *p,p'*-DDT is fed to rats, *p,p'*-DDT accumulates. Our data are completely consistent with the view that the supposed isomeric conversion does not occur. The original assumptions of Klein *et al.* (*1, 2*) are based on faulty data concerning the purity of *o,p'*-DDT, and the major conclusion is incorrect.

The correct explanation for the rapid disappearance of *o,p'*-DDT when it is fed and for the appearance of *p,p'*-DDT can now be given. *o,p'*-DDT, the major constituent of commercial

o,p'-DDT preparations, by virtue of the open positions on the ring bearing the *o*-Cl, is rapidly converted to hydroxy and methoxy metabolites that are rapidly excreted in the feces (*11*). In contrast, *p,p'*-DDT, the minor impurity, is relatively inert metabolically and accumulates in the lipid of the animal body.

Thus, this differential metabolic behavior leads to the simultaneous disappearance of the major component of relatively pure samples of *o,p'*-DDT, with the concomitant appearance of the minor constituent, the *p,p'*-DDT impurity. There is no existing chemical or biological information or data to support the idea that *o,p'*-DDT is converted to the *p,p'*-DDT isomer.

References

1. A. K. Klein, *et al.*, *J. Assoc. Offic. Anal. Chem.* **47**, 1129 (1964).
2. A. K. Klein, E. P. Laug, P. R. Datta, J. L. Mendel, *J. Am. Chem. Soc.* **87**, 2520 (1965).
3. Abbreviations are as follows: *p,p'*-DDT is 1,1,1-trichloro-2,2-bis(*p*-chlorophenyl)ethane; *o,p'*-DDT is 1,1,1-trichloro-2-(*p*-chlorophenyl)-2-(*o*-chlorophenyl)ethane; *p,p'*-DDE is 1,1-dichloro-2,2-bis(*p*-chlorophenyl)ethylene; *o,p'*-DDE is 1,1-dichloro-2-(*p*-chlorophenyl)-2-(*o*-chlorophenyl)ethylene; *o,p'*-DDD, 1,1,-dichloro-2-(*o*-chlorophenyl)-2-(*p*-chlorophenyl)ethane.
4. E. S. Gould, *Mechanism and Structure in Organic Chemistry* (Holt, Rinehart and Winston, New York, 1959); L. L. Ingraham, *Biochemical Mechanisms* (Wiley, New York, 1962).
5. J. Bitman, H. C. Cecil, S. J. Harris, G. F. Fries, *J. Agr. Food Chem.* **19**, 333 (1971).
6. H. L. Haller *et al.*, *J. Am. Chem. Soc.* **67**, 1591 (1945).
7. D. J. Ecobichon and P. W. Saschenbrecker, *Can. J. Physiol. Pharmacol.* **46**, 785 (1968).
8. M. C. French and D. J. Jefferies, *Science* **165**, 914 (1969).
9. J. L. Mendel, A. K. Klein, J. T. Chen, M. S. Walton, *J. Assoc. Offic. Anal. Chem.* **50**, 897 (1967).
10. T. G. Lamont, G. E. Bagley, W. L. Reichel, *Bull. Environ. Contam. Toxicol.* **5**, 231 (1970).
11. V. J. Feil, E. J. Thacker, R. G. Zaylskie, H. Lamoureux, E. Styrvoky, Abstract, 162nd Meeting, American Chemical Society, 7–12 September 1971, Washington, D.C.

Effect of Phenobarbital Pretreatment on the Metabolism of DDT in the Rat and the Bovine[1,2]

JEAN-GUY ALARY, PATRICK GUAY, AND JULES BRODEUR

The persistence of organochlorine insecticide residues in animal tissues intended for human food consumption raises the problem of effecting their removal from the live animal. During the past few years, various groups of workers have carried out investigations aimed at achieving a more rapid mobilization and excretion of insecticide residues from animal tissues. According to Wesley *et al.* (1966, 1969), starvation accelerates the depletion rate of DDT residues from abdominal fat samples and egg yolk of laying hens. Similarly, Donaldson *et al.* (1968) presented evidence suggesting that cyclic starvation increases the rate of removal of DDT residues in chick tissues. Miller (1967) was able to show that the administration of thyroprotein appreciably lowers the elimination of DDT residues in milk fat of dairy cattle fed a hypocaloric diet. Stull *et al.*

[1] Abbreviations used: DDT or p,p'-DDT, 1,1,1-trichloro-2,2-bis(p-chlorophenyl)ethane; DDD or p,p'-DDD, 1,1-dichloro-2,2-bis(p-chlorophenyl)ethane; DDE or p,p'-DDE, 1,1-dichloro-2,2-bis-(p-chlorophenyl)ethylene; DDA, bis(p-chlorophenyl)acetic acid; parathion, O,O-diethyl O-(4-nitro-phenyl) phosphorothionate; paraoxon, O,O-diethyl O-(4-nitrophenyl) phosphate.

[2] This work was supported by grants from the Medical Research Council of Canada (MA-1938) and Le Conseil de Recherches Agricoles, Province de Québec (MM-66-271).

150

(1968) demonstrated a slight but transitory increase in the daily excretion of DDT in milk, following treatment of dairy cows with thyroprotein. Similarly, treatment of beef cattle with a mixture of vitamins A, D, and E appeared to reduce somewhat the concentration of dieldrin residues in adipose tissue (Hironaka, 1968).

The problem of pesticide residue mobilization and elimination has also been studied by various groups of workers using a quite different approach. Thus, stimulation of microsomal oxidizing enzyme systems in the liver of rats by pretreatment with phenobarbital was followed by an increased urinary excretion of benzene hexachloride metabolites (Koransky et al., 1964). Similarly, pretreatment of various groups of experimental animals with DDT, heptabarbital, aminopyrine, tolbutamide, or phenylbutazone was effective in reducing the storage of dieldrin and heptachlor in the adipose tissue and in accelerating the urinary and fecal excretion of polar metabolites of dieldrin (Street, 1964; Street and Blau, 1966; Street et al., 1966a,b; Street and Chadwick, 1967; Cueto and Hayes, 1967). On the other hand, Deichmann et al. (1969) were unable to reproduce these results in dogs fed a mixture of aldrin and DDT. Recently, Davies et al. (1969) reported that blood concentrations of DDE in human subjects taking diphenylhydantoin and phenobarbital were considerably reduced in comparison with those of healthy controls not taking these drugs. The present investigation was therefore undertaken to evaluate the effect of phenobarbital, a potent hepatic microsomal enzyme inducer, on the excretion pattern of the fat-soluble metabolites of DDT in milk of dairy cows and of water-soluble metabolite DDA in urine of beef cattle and rats. The effect of phenobarbital was also studied in vitro on the metabolism of DDT by liver homogenates of rats. The results obtained tend to support the hypothesis that phenobarbital acts by stimulating the hepatic biotransformation of DDT. A preliminary account of this study has been published elsewhere (Alary et al., 1968).

MATERIALS AND METHODS

In the first series of experiments, 4 lactating Holstein cows (average weight 365 kg) were used. Each animal was given 50 mg of DDT (99% pure) in the diet twice daily for 63 consecutive days. Stock solutions of DDT were prepared daily by dissolving 1 g of purified DDT in 5 ml of acetone; 0.25 ml portions of this stock solution were then added dropwise to a small quantity of powdered feed. The animals were first given the treated feed, and the normal feed was then made available ad libitum. From the day 28 until the day 49, inclusive, the cows received daily ip injections of 5 mg/kg phenobarbital sodium in a 5% sterile dextrose vehicle. Throughout the experiment, the cows were milked twice daily and the milk from two consecutive milkings, 1600 and 0700 hr, was collected at various intervals ranging from 2 to 5 days, pooled, measured, and kept at ±4°C in glass bottles pending further determinations. The extraction of fat from 150 g of whole milk and the analysis for chlorinated insecticide residues of DDT type by electron capture GLC on an Aerograph model 2100 gas chromatograph were carried out according to the procedure of McCully and McKinley (1964). A U-shaped Pyrex glass column (1.8 m × 0.64 cm) containing 4% SE-30 and 6% QF-1 silicones on acid-washed Chromosorb W (60/80 mesh) was used. Optimal temperatures (°C) were 190 (column), 210 (detector), and 220 (injector). Optimal nitrogen flow rate was 80 ml/min. Results were expressed in terms of ppm of pesticide residues in whole milk.

151

In a second series of studies, 6 young bulls weighing between 320 and 360 kg were used. In order to facilitate quantitative urine collection, a uretrotomy was performed on the animals, under local anesthesia, according to the procedure described by Guay et al. (1968). The animals were then placed in 2 groups of 3 animals each; in 2 animals of each group, phenobarbital sodium in a 5% sterile dextrose vehicle was injected ip at a dose of 5 mg/kg/day, the treatment being administered for 7 and 14 days in the first and second group, respectively. The third animal in each group served as a control and received an equivalent volume of the vehicle (0.05 ml/kg/day). Twenty-four hours after the last injection of phenobarbital, all six animals received a single iv injection of 2 g DDT (99% pure) dissolved in 6 ml of acetone. Twenty-four-hour urine samples were collected quantitatively at various time intervals before and after the administration of DDT, for measurement of the water-soluble metabolite DDA, 200-ml aliquots of urine being used for this purpose. After low speed centrifugation for 5 min to remove solid materials, the urine was acidified at pH 2 with 6 N sulfuric acid. DDA was then extracted essentially according to the general procedure of Pinto et al. (1965). Thus, 200 ml of acidified urine were extracted during 4 min with an equal volume of ether by shaking in a separatory funnel. After separation, the ether layer was concentrated to a volume of 100 ml and washed during 1 min with 100 ml of 3 N sulfuric acid; 2 further washings were carried out during 1 min each, with 50 ml of 5% sodium chloride. The ether was then extracted with 100 ml of 0.5 N sodium hydroxide during 4 min in a separatory funnel. The aqueous layer was acidified to pH 1.5 with concentrated hydrochloric acid, and reextracted with 100 ml of chloroform by shaking in a separatory funnel during 4 min. Chloroform was evaporated until a dry residue was obtained in a Büchner flask evaporator. The DDA thus extracted was measured according to the colorimetric method of Ofner and Calvery (1945). The colored material resulting from this final step possessed an absorption spectrum in the visible range identical to that of standard DDA similarly extracted and colored.

In the two bulls pretreated during 14 days, liver biopsy, according to the procedure of Holtenius (1961), was performed 24 hr after the last injection of phenobarbital; the liver of both control animals was also biopsied at the same time. Portions of all four biopsy materials were then placed in ice-cold Krebs-Ringer bicarbonate buffer, pH 7.4, for immediate determination of the enzymatic activity of parathion oxidase and paraoxonase according to the method described by Alary and Brodeur (1969). These enzymes are involved in the oxidative metabolism of parathion and the hydrolytic metabolism of paraoxon, respectively.

In the third series of experiments, 59 adult female Holtzman rats (180–200 g) were used. Of these, 45 were divided into 9 equal groups and treated as follows: 15 animals received phenobarbital sodium ip at a daily dose of 50 mg/kg during 5 days, 15 received the same amount of phenobarbital during 25 days, and 15 were given saline only for 25 days and served as controls. At the end of the pretreatment period, all animals received a single ip injection of 100 mg/kg DDT (99% pure) dissolved in corn oil. Forty-eight hours before receiving DDT, the animals were placed in groups of 2 or 3 in metabolism cages for quantitative urine collection and determination of urinary DDA at various times after DDT administration. The urine of each group of 5 animals was pooled and made up to a volume of 20 ml with normal saline. DDA was extracted and determined according to the procedure previously described.

Finally, the rate of *in vitro* metabolism of DDT was measured in rat liver homogenates. For this purpose, 14 adult female rats were distributed into 2 groups of 7 animals each; the first group received phenobarbital ip at a dose of 50 mg/kg/day for 5 days, and the other group was given normal saline only for the same period. Twenty-four hours after the last injection of phenobarbital, the animals of both groups were sacrificed; their livers were removed, blotted, weighed, and homogenized in ice-cold Krebs-Ringer bicarbonate buffer, pH 7.4. A 1-ml portion of a 20% liver homogenate (200 mg of tissue) was then added to a test-system containing: 0.4 ml of NADP (8 mg/ml), 0.2 ml of glucose 6-phosphate (20 mg/ml), 0.3 ml of nicotinamide (0.16 mg/ml), 0.1 ml MgCl$_2$ (20 mg/ml), 0.1 ml of recrystallized *p,p'*-DDT (1 mg/ml), and a sufficient amount of Krebs-Ringer bicarbonate buffer to make a final volume of 3.0 ml. Incubation was carried out at 37°C for 120 min in a Dubnoff shaking incubator under aerobic conditions. The reaction was stopped by addition of 6 ml of chloroform and 0.5 ml 6 N HCl (Morello, 1965). After shaking in a separatory funnel, the chloroform layer was separated; two additional extractions with chloroform were carried out, followed by evaporation to dryness of the combined extracts under a stream of air. From then on, quantitative extraction of DDT and its metabolites was performed according to the procedure described by Saschenbrecker and Ecobichon (1967). Quantitative evaluations of *p,p'*-DDT, *p,p'*-DDD, and *p,p'*-DDE were carried out on an Aerograph model 2100 gas chromatograph, equipped with an electron capture detector, as specified above.

RESULTS

Effect of Phenobarbital on the Excretion of DDT and Fat-Soluble Metabolites in Cow's Milk

During the period of the experiment, each of the 4 cows received 6.3 g of DDT divided into daily doses of 100 mg. The animals continued to gain weight at a normal rate and showed no signs of intoxication. However, light sedation was noted in all 4 animals during treatment with phenobarbital. Figure 1 shows the effect of phenobarbital treatment on the milk excretion of fat-soluble residues of DDT (*p,p'*-DDT, *p,p'*-DDD,

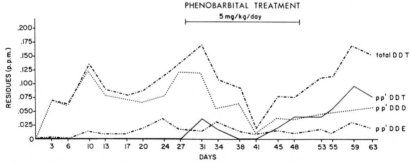

Fig. 1. Effect of phenobarbital pretreatment on the excretion of total DDT-derived materials (sum of *p,p'*-DDT, *p,p'*-DDD, and *p,p'*-DDE), *p,p'*-DDT, *p,p'*-DDD, and *p,p'*-DDE in milk of dairy cows. The curves represent the means of determinations made on milk samples from 4 animals at the time intervals indicated on the abscissa.

p,p'-DDE), referred to as total DDT. It can be seen that, during the period of treatment with the barbiturate, a decrease in total DDT occurred. A regression line drawn from the data collected during the entire period of treatment with phenobarbital showed that the concentration of total DDT metabolites correlates negatively with time ($r = -0.4962$ $P < 0.02$). A significant reduction in the excretion of p,p'-DDD (Fig. 1) appears to be responsible for this effect, since the concentration of this metabolite also presented a negative correlation with time ($r = -0.5583$, $P < 0.01$). From Fig. 1, it is also seen that the excretion pattern of p,p'-DDE was not altered by the administration of phenobarbital, while the appearance of p,p'-DDT in milk was coincident with the initiation of the phenobarbital treatment. At the end of the treatment period and during the following days, p,p'-DDT was a major excretion product and would seem to be responsible for the large increase in total DDT material excreted at that time. It should be pointed out that no noticeable changes occurred in total milk yield and milk fat content throughout the study.

Effect of Phenobarbital on the Urinary Excretion of DDA in Young Bulls and Adult Female Rats

Pretreatment of young bulls with phenobarbital at a daily dose of 5 mg/kg during 7 or 14 days, followed by the iv administration of 2 g of DDT, resulted in a marked increase in the rate of urinary excretion of the water-soluble metabolite DDA. Thus, as

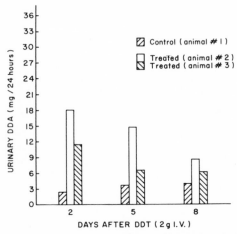

FIG. 2. Effect of phenobarbital pretreatment (5 mg/kg/day, for 7 days) on the 24-hr urinary excretion of DDA in young bulls given DDT; 3 animals were used: 1 control and 2 treated. Each bar represents the value of a single determination made on each animal.

shown in Figs. 2 and 3, 2 days after the administration of DDT, all animals given phenobarbital for 7 or 14 days showed very high levels of DDA in the urine as compared with their respective controls. At 5 and 8 days after DDT, a tendency toward a higher excretion of DDA was still noted in some of the treated animals, whereas this pattern had disappeared at 15 and 22 days after DDT in the group given phenobarbital over a period of 14 days.

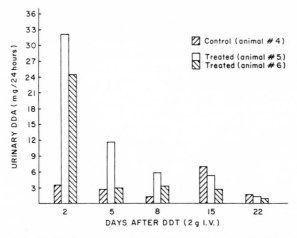

Fig. 3. Effect of phenobarbital pretreatment (5 mg/kg/day, for 14 days) on the 24-hr urinary excretion of DDA in young bulls given DDT; 3 animals were used: 1 control and 2 treated. Each bar represents the value of a single determination made on each animal.

Similarly, as shown in Fig. 4, adult female rats pretreated with phenobarbital at a daily dose of 50 mg/kg excreted significantly larger amounts of DDA in the urine than untreated animals during the first 24 hr after DDT. During the period ranging from 24 to 72 hr, no significant difference was noted between the treated and control groups. Between 72 and 120 hr, however, controls excreted significantly more DDA than any of the treated animals. Data showing cumulative excretion of DDA over the entire period of observation indicate that control rats eliminated an average of 114.4 μg of the metabolite as compared with 152.4 and 133.5 μg for animals treated with phenobarbital during 5 and 25 days, respectively. Further determinations showed that elimination of

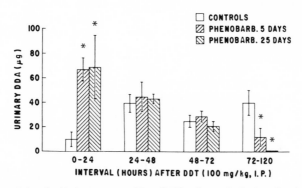

Fig. 4. Effect of phenobarbital pretreatment (50 mg/kg/day, for 5 or 25 days) on the urinary excretion of DDA in adult female rats given DDT. Each bar represents the mean ± SE of determinations made on the urine of 3 groups of 5 animals. The asterisk indicates that the amount of DDA was significantly different ($p < 0.05$) from controls.

155

DDA persisted for as long as 14 days after DDT in the controls, whereas no DDA could be detected 5 days after DDT in the treated animals.

Effect of Phenobarbital on the in Vitro Metabolism of DDT and Parathion by Liver Homogenates

The results of the experiments in rats are shown in Table 1. It can be seen, first, that pretreatment of the animals with phenobarbital gave rise to a 2-fold increase in the amount of p,p'-DDD present in the medium at the end of the incubation period, while no change occurred in the case of p,p'-DDE. A marked and significant decrease in the amount of residual p,p'-DDT was observed upon incubation of treated livers with the chlorinated hydrocarbon. Thus, the sum of total fat-soluble material recovered from the incubation medium was 31.9 μg in treated animals as compared with 55.6 μg in the controls. Although part of the unaccounted material was lost during the extraction procedure (approximately 15% with p,p'-DDT and 20–25% with p,p'-DDD), these results further suggest that phenobarbital accelerated the transformation of DDT to metabolites which could not be detected by the present method of analysis.

TABLE 1

EFFECT OF PHENOBARBITAL[a] ON THE *in vitro* METABOLISM OF 100 μG p,p'-DDT BY LIVER HOMOGENATES OF ADULT FEMALE RATS

| Treatment | Amount of chlorinated hydrocarbon (μg)[b] | | |
	p,p'-DDT	p,p'-DDD	p,p'-DDE
None	49.9 ± 2.2	5.1 ± 0.2	0.6 ± 0.05
Phenobarbital	20.8 ± 2.3^c	10.5 ± 1.0^c	0.6 ± 0.05

[a] 50 mg/kg/day for 5 days.
[b] Each value represents the mean ± SE of 7 determinations.
[c] Differs from untreated group, $p < 0.001$.

The results of the study in bulls are shown in Tables 2 and 3. As can be seen, pretreatment with phenobarbital considerably stimulated the activity of the enzyme systems that catalyze the transformation of parathion and paraoxon. Due to the very limited amount of tissue obtained at biopsy, it was not possible to measure the *in vitro* metabolic activity of the bovine liver against DDT.

TABLE 2

EFFECT OF PHENOBARBITAL[a] ON THE ENZYME SYSTEM RESPONSIBLE FOR THE OXIDATION OF PARATHION TO PARAOXON BY BOVINE LIVER

| Treatment | Enzyme activity[b] (μg paraoxon formed/10 mg liver/10 min) | |
	Before treatment	After treatment
None	0.51	0.64
None	—[c]	0.56
Phenobarbital	0.56	>1.00
Phenobarbital	0.41	>1.00

[a] 5 mg/kg/day for 14 days.
[b] Enzyme activity was determined according to Alary and Brodeur (1969).
[c] No tissue available.

TABLE 3
EFFECT OF PHENOBARBITAL[a] ON THE PARAOXONASE
ACTIVITY OF BOVINE LIVER

Treatment	Enzyme activity[b] after treatment (μg p-nitrophenol formed/10 mg tissue/10 min)
None	6.4
None	6.0
Phenobarbital	11.0
Phenobarbital	11.2

[a] 5 mg/kg/day for 14 days.
[b] Enzyme activity was determined according to Alary and Brodeur (1969).

DISCUSSION

Repeated administration of low doses of phenobarbital sodium to dairy cows given DDT for 63 days resulted in a significant decrease in the content of total DDT metabolites in the milk. Pretreatment with phenobarbital sodium also markedly increased the urinary excretion of DDA in bulls administered a single dose of DDT. A possible explanation for these observations is that phenobarbital acts by stimulating the activity of the liver microsomal enzymes involved in the biotransformation of DDT, as already suggested by various groups of workers using different combinations of inducers and chlorinated hydrocarbon insecticides (Koransky et al., 1964; Street, 1964; Street and Blau, 1966; Street et al., 1966a,b; Street and Chadwick, 1967; Cueto and Hayes, 1967). This hypothesis appears to be substantiated indirectly by data presented in this paper indicating that, in bulls pretreated with phenobarbital at a daily dose of 5 mg/kg during 14 days, enzyme induction was actually present, as shown by an increased rate of metabolism of parathion and paraoxon under in vitro conditions by liver homogenates. Parathion and paraoxon are known to undergo oxidative and hydrolytic metabolism, respectively, in a series of reactions catalyzed by microsomal enzymes (Neal and DuBois, 1965; Nakatsugawa and Dahm, 1967). Bovine liver, therefore, like that of several other mammalian species, appears to be susceptible to enzyme induction. More direct evidence for the inducing effect of phenobarbital on the metabolism of DDT has been derived from in vitro studies showing that the barbiturate significantly increases the metabolic transformation of p,p'-DDT to p,p'-DDD by liver homogenates of rats, and from morphologic observations showing a marked proliferation of smooth endoplasmic reticulum in the liver cells of the bovines used in this study (Côté et al., personal communication).

Understanding of the mechanism by which phenobarbital produces its effect on DDT metabolism is facilitated by consideration of the biotransformation pathways of DDT. The following sequence has been suggested by Peterson and Robison (1964) as most likely representing the possible metabolic fate of DDT in the rat[3]:

DDE

DDT

DDD → DDMU → DDMS → DDNU →
→ DDOH → probable intermediate aldehyde → DDA

[3] Abbreviations: DDMU, 1-chloro-2,2-bis(p-chlorophenyl)ethylene; DDMS, 1-chloro-2,2-bis-(p-chlorophenyl)ethane; DDNU, unsym-bis(p-chlorophenyl)ethylene; DDOH, 2,2-bis(p-chloro-phenyl)ethanol.

This series of metabolic steps appears to be common to other organisms, since a similar sequence has also been proposed in rabbits (Bowery et al., 1965), birds (Abou-Donia and Menzel, 1968), and bacteria (Wedemeyer, 1967). Although extensive metabolic studies have not been conducted in bovines, investigations by McCully et al. (1966), Fries and Kane (1967), and Whiting et al. (1968) have provided sufficient evidence to indicate that DDT undergoes metabolic transformation along similar lines. However, the site of formation of DDD or DDE from DDT in vertebrates has been questioned by several authors. There is conclusive evidence that part of the DDD formed originates from microbial action in the rumen or the intestine (Barker et al., 1965; Miskus et al., 1965, Mendel and Walton, 1966); it is not certain whether the conversion of DDT to DDE can also be accomplished by the rumen (Witt et al., 1966). In addition, Castro (1964) and Miskus et al. (1965) observed a transformation of DDT to DDD by dilute solutions of reduced porphyrin complexes. The nonenzymatic breakdown of DDT to DDD and/or DDE reported by Ecobichon and Saschenbrecker (1967) and Whiting et al. (1968) was probably due to catalytic action of iron porphyrins from damaged blood cells. Despite the previous observations, there is sufficient direct and indirect evidence in support of the view that the liver is an important site for the enzymatically catalyzed transformation of DDT to DDD in the rat (Datta et al., 1964; Klein et al., 1964; Peterson and Robison, 1964; Vessey et al., 1968), the bird (Bailey et al., 1969), and the bovine (Fries and Kane, 1967; Whiting et al., 1968). In addition, Morello (1965) and Sanchez (1967) have shown that the conversion of DDT to DDD occurs in the microsomal fraction of the liver cell. On the other hand, the liver does not appear to be the site of formation of DDE from DDT, either in the rat (Peterson and Robison, 1964) or in the bovine (Whiting et al., 1968). It is now accepted that DDE is not an intermediate step in the conversion of DDT to DDD and its final water-soluble metabolite DDA in various species (Peterson and Robison, 1964; Fries and Kane, 1967; Abou-Donia and Menzel, 1968). Although DDE is a urinary metabolite of DDT (Cueto and Biros, 1967; Laws et al., 1967), it now appears that DDE can be slowly degraded further, at least in birds (Abou-Donia and Menzel, 1968; Ecobichon and Saschenbrecker, 1968).

The evidence accumulated during this investigation strongly suggests that phenobarbital stimulates the enzymatic biotransformation of DDT to DDD in the liver. This is supported directly by the in vitro data obtained after incubation of DDT with homogenates of liver taken from control and phenobarbital-pretreated rats. There is no evidence in the literature to suggest that further metabolism of DDD to DDA actually takes place in the liver. It is not known, therefore, whether or not this sequence of metabolic reactions can also be stimulated by phenobarbital. It is logical to postulate, however, that the formation of DDA would at least keep pace with that of DDD, and, in the final analysis, that microsomal enzyme induction would result in an increased production of DDA, as observed in this study. The fact that the concentrations of p,p'-DDD in milk undergo a marked and significant decrease after administration of phenobarbital tends to support this hypothesis.

The body of evidence accumulated thus far suggests that the small amounts of DDE formed during incubation of liver homogenates with DDT originated from the catalytic action of contaminating hemolyzed red blood cells (Ecobichon and Saschenbrecker, 1967; Whiting et al., 1968). The observation that phenobarbital did not influence the rate of formation of DDE under these *in vitro* conditions, and that the excretion of DDE in milk was not affected by pretreatment with phenobarbital, seems to provide further evidence that the liver is not involved in the metabolic transformation of DDT to DDE. It appears therefore that the rate of excretion of *p,p'*-DDE in milk is in equilibrium with the rate of formation of this metabolite at extrahepatic sites.

It is difficult to explain why the excretion of measurable amounts of *p,p'*-DDT in milk coincides with the initiation of phenobarbital administration. A possible explanation might be that, despite an increased biotransformation of DDT upon the influence of the inducer, a slow and progressive build-up of the unmetabolized pesticide occurs in the adipose tissue: thus, the presence of *p,p'*-DDT in milk would represent a state of saturation of the adipose tissue by the pesticide.

An alternative explanation to the general effect of phenobarbital on the metabolism of DDT could be that phenobarbital initiates nonspecific mobilization of pesticide residues from the adipose tissue, either by displacement of the residues themselves or by some other unknown mechanisms. If this were true, then one would expect to find a parallel increase of the three fat-soluble residues in milk; the results presented here demonstrate that this is not the case as *p,p'*-DDT, *p,p'*-DDD, and *p,p'*-DDE undergo very different changes in their concentrations under the influence of phenobarbital.

The body of evidence accumulated in this and other investigations strongly supports the hypothesis that phenobarbital stimulates the metabolism of DDT by the liver microsomal enzymes. The net result seems to be a mobilization of DDT-derived materials from adipose tissue and a reduction of the total body burden of the pesticide.

ACKNOWLEDGMENTS

The authors acknowledge the assistance of Mr. G. Léveillé in the determination of pesticide residues in milk.

REFERENCES

ABOU-DONIA, M. B., and MENZEL, D. B. (1968). The metabolism in vivo of 1,1-trichloro-2,2-bis (*p*-chlorophenyl) ethane (DDT), 1,1-dichloro-2,2-bis (*p*-chlorophenyl) ethane (DDD) and 1,1-dichloro-2,2-bis (*p*-chlorophenyl) ethylene (DDE) in the chick by embryonic injection and dietary ingestion. *Biochem. Pharmacol.* **17**, 2143–2161.

ALARY, J.-G., and BRODEUR, J. (1969). Studies on the mechanism of phenobarbital-induced protection against parathion in adult female rats. *J. Pharmacol. Exp. Ther.* **169**, 159–167.

ALARY, J.-G., BRODEUR, J., COTE, M. G., PANISSET, J. C., LAMOTHE, P., and GUAY, P. (1968). The effect of a pretreatment with phenobarbital on the metabolism of DDT in dairy cows and young bulls. *Rev. Can. Biol.* **27**, 269–271.

BAILEY, S., BUNYAN, P. J., RENNISON, B. D., and TAYLOR, A. (1969). The metabolism of 1,1-di-(*p*-chlorophenyl)-2,2,2-trichloroethane and 1,1-di(*p*-chlorophenyl)-2,2-dichloroethane in the pigeon. *Toxicol. Appl. Pharmacol.* **14**, 13–22.

BARKER, P. S., MORRISON, F. O., and WHITAKER, R. S. (1965). Conversion of DDT to DDD by *Proteus vulgaris*, a bacterium isolated from the intestinal flora of a mouse. *Nature (London)* **205**, 621–622.

BOWERY, T. G., GATTERDAM, P. E., GUTHRIE, F. E., and RABB, R. L. (1965). Fate of inhaled C^{14}-TDE in rabbits. *J. Agr. Food Chem.* **13**, 356–359.

CASTRO, C. E. (1964). The rapid oxidation of iron (II) porphyrins by alkyl halides: a possible mode of intoxication of organisms by alkyl halides. *J. Amer. Chem. Soc.* **86**, 2310–2311.

CUETO, C., and BIROS, F. J. (1967). Chlorinated insecticides and related materials in human urine. *Toxicol. Appl. Pharmacol.* **10**, 261–269.

CUETO, C., and HAYES, W. J. (1967). Effect of repeated administration of phenobarbital on the metabolism of dieldrin. *Ind. Med. Surg.* **36**, 546–551.

DATTA, P. R., LAUG, E. P., and KLEIN, A. K. (1964). Conversion of *p,p'*-DDT to *p,p'*-DDD in the liver of the rat. *Science* **145**, 1052–1053.

DAVIES, J. E., EDMUNDSON, W. F., CARTER, C. H., and BARQUET, A. (1969). Effect of anticonvulsant drugs on dicophane (D.D.T.) residues in man. *Lancet* **2**, 7–9.

DEICHMANN, W. B., KEPLINGER, M., DRESSLER, I., and SALA, F. (1969). Retention of dieldrin and DDT in the tissues of dogs fed aldrin and DDT individually and as a mixture. *Toxicol. Appl. Pharmacol.* **14**, 205–213.

DONALDSON, W. E., SHEETS, T. J., and JACKSON, M. D. (1968). Starvation effects on DDT residues in chick tissues. *Poultry Sci.* **47**, 237–243.

ECOBICHON, D. J., and SASCHENBRECKER, P. W. (1967). Dechlorination of DDT in frozen blood. *Science* **156**, 663–665.

ECOBICHON, D. J., and SASCHENBRECKER, P. W. (1968). Pharmacodynamic study of DDT in cockerels. *Can. J. Physiol. Pharmacol.* **46**, 785–794.

FRIES, G. F., and KANE, E. A. (1967). Retention of DDT and DDE by the bovine. *J. Dairy Sci.* **50**, 1512–1515.

GUAY, P., LAMOTHE, P., ALARY, J.-G., BRODEUR, J., CÔTÉ, M. G., and PANISSET, J. C. (1968). L'urétrotomie expérimentale chez le taureau. *Econ. Med. Anim.* **9**, 19–25.

HIRONAKA, R. (1968). Elimination of dieldrin from beef cattle. *Can. Vet. J.* **9**, 167–169.

HOLTENIUS, P. (1961). Cytological puncture. A new method for the study of bovine hepatic disease. *Cornell Vet.* **51**, 56–63.

KLEIN, A. K., LAUG, E. P., DATTA, P. R., WATTS, J. O., and CHEN, J. T. (1964). Metabolites: reductive dechlorination of DDT to DDD and isomeric transformation of *o,p'*-DDT to *p,p'*-DDT in vivo. *J. Ass. Offic. Agr. Chemists* **47**, 1129–1145.

KORANSKY, W., PORTIG, J., VOHLAND, H. W., and KLEMPAU, I. (1964). Die Elimination von α- und β-Hexachlorcyclohexan und irhe Beeinflussung durch Enzyme der Lebermikrosomen. *Naunyn-Schmiedebergs Arch. Exp. Pathol. Pharmakol.* **247**, 49–60.

LAWS, E. R., CURLEY, A., and BIROS, F. J. (1967). Men with intensive occupational exposure to DDT. A clinical and chemical study. *Arch. Environ. Health* **15**, 766–775.

MCCULLY, K. A., and MCKINLEY, W. P. (1964). Determination of chlorinated pesticide residues in fat by electron capture gas chromatography. *J. Ass. Offic. Agr. Chemists* **47**, 652–659.

MCCULLY, K. A., VILLENEUVE, D. C., and MCKINLEY, W. P. (1966). Metabolism and storage of DDT in beef cattle. *J. Ass. Offic. Agr. Chemists* **49**, 966–973.

MENDEL, J. L., and WALTON, M. S. (1966). Conversion of *p,p'*-DDT to *p,p'*-DDD by intestinal flora of the rat. *Science* **151**, 1527–1528.

MILLER, D. D. (1967). Effect of thyroprotein and a low-energy ration on removal of DDT from lactating dairy cows. *J. Dairy Sci.* **50**, 1444–1447.

MISKUS, R. P., BLAIR, D. P., and CASIDA, J. E. (1965). Conversion of DDT to DDD by bovine rumen fluid, lake water, and reduced porphyrins. *J. Agr. Food Chem.* **13**, 481–483.

MORELLO, A. (1965). Induction of DDT-metabolizing enzymes in microsomes of rat liver after administration of DDT. *Can. J. Biochem.* **43**, 1289–1293.

NAKATSUGAWA, T., and DAHM, P. A. (1967). Microsomal metabolism of parathion. *Biochem. Pharmacol.* **16**, 25–38.

NEAL, R. A., and DUBOIS, K. P. (1965). Studies on the mechanism of detoxification of cholinergic phosphorothioates. *J. Pharmacol. Exp. Ther.* **148**, 185–192.

OFNER, R. R., and CALVERY, H. O. (1945). Determination of DDT (2,2-bis (*p*-chlorophenyl) 1,1,1-trichloroethane) and its metabolite in biological materials by use of the Schechter-Haller method. *J. Pharmacol. Exp. Ther.* **85**, 363–370.

PETERSON, J. E., and ROBISON, W. H. (1964). Metabolic products of *p,p'*-DDT in the rat. *Toxicol. Appl. Pharmacol.* **6**, 321–327.

PINTO, J. D., CAMIEN, M. N., and DUNN, M. S. (1965). Metabolic fate of *p,p'*-DDT [1,1,1-trichloro-2,2-bis (*p*-chlorophenyl) ethane] in rats. *J. Biol. Chem.* **240**, 2148–2154.

SANCHEZ, E. (1967). DDT-induced metabolic changes in rat liver. *Can. J. Biochem.* **45**, 1809–1817.

SASCHENBRECKER, P. W., and ECOBICHON, D. J. (1967). Extraction and gas chromatographic analysis of chlorinated insecticides from animal tissues. *J. Agr. Food Chem.* **15**, 168–170.

STREET, J. C. (1964). DDT antagonism to dieldrin storage in adipose tissue of rats. *Science* **146**, 1580–1581.

STREET, J. C., and BLAU, A. D. (1966). Insecticide interactions affecting residue accumulation in animal tissues. *Toxicol. Appl. Pharmacol.* **8**, 497–504.

STREET, J. C., and CHADWICK, R. W. (1967). Stimulation of dieldrin metabolism by DDT. *Toxicol. Appl. Pharmacol.* **11**, 68–71.

STREET, J. C., CHADWICK, R. W., WANG, M., and PHILLIPS, R. L. (1966a). Insecticide interactions affecting residue storage in animal tissues. *J. Agr. Food Chem.* **14**, 545–549.

STREET, J. C., WANG, M., and BLAU, A. D. (1966b). Drug effects on dieldrin storage in rat tissue. *Bull. Environ. Contamination Toxicol.* **1**, 6–15.

STULL, J. W., BROWN, W. H., WHITING, F. M., SULLIVAN, L. M., MILBRATH, M., and WITT, J. M. (1968). Secretion of DDT by lactating cows fed thyroprotein. *J. Dairy Sci.* **51**, 56–59.

VESSEY, D. A., MAYNARD, L. S., BROWN, W. H., and STULL, J. W. (1968). The subcellular partition and metabolism of orally administered 1,1,1-trichloro-2,2-bis (*p*-chlorophenyl) ethane by rat liver cells. *Biochem. Pharmacol.* **17**, 171–174.

WEDEMEYER, G. (1967). Dechlorination of 1,1,1-trichloro-2,2-bis (*p*-chlorophenyl) ethane by *Aerobacter aerogenes*. I. Metabolic Products. *Appl. Microbiol.* **15**, 569–574.

WESLEY, R. L., STEMP, A. R., and STADELMAN, W. J. (1966). Depletion of DDT from commercial layers. *Poultry Sci.* **45**, 321–324.

WESLEY, R. L., STEMP, A. R., HARRINGTON, R. B., LISKA, B. J., ADAMS, R. L., and STADELMAN, W. J. (1969). Further studies on depletion of DDT residues from laying hens. *Poultry Sci.* **48**, 1269–1275.

WHITING, F. M., HAGYARD, S. B., BROWN, W. H., and STULL, J. W. (1968). Detoxification of DDT by the perfused bovine liver. *J. Dairy Sci.* **51**, 1612–1615.

WITT, J. M., WHITING, F. M., BROWN, W. H., STULL, J. W. (1966). Contamination of milk from different routes of animal exposure to DDT. *J. Dairy Sci.* **49**, 370–380.

161

DDT Metabolism: Oxidation of the Metabolite 2,2-bis(p-Chlorophenyl)ethanol by Alcohol Dehydrogenase

Joseph E. Suggs, Robert E. Hawk
August Curley
Elizabeth L. Boozer
James D. McKinney

The metabolism of DDT [1,1,1-trichloro-2,2-bis(p-chlorophenyl)ethane] in mammals gives rise to 2,2-bis(p-chlorophenyl)ethanol (DDOH) and 2,2-bis(p-chlorophenyl) acetic acid (DDA) (1). An intermediate aldehyde has been proposed but has not yet been found to occur in vivo. We have synthesized the proposed intermediate, 2,2-bis(p-chlorophenyl)acetaldehyde (DDCHO) (2).

Because the aldehyde has never been found in vivo and because preliminary studies indicate that the synthetic compound is highly unstable and reactive, the aldehyde was examined as a possible product of oxidation of DDOH by crystalline liver alcohol dehydrogenase (E.C.1.1.1.1).

The oxidation of DDOH was detected in a double-beam spectrophotometer by following the reduction of nicotinamide-adenine dinucleotide (NAD) at 340 nm in the presence of crystalline horse liver alcohol dehydrogenase (3). Because DDOH is insoluble in aqueous media, the compound was dissolved in 50 percent glycerolformal before its addition to the buffered incubation media. The resulting cloudy suspension prevented accurate spectrophotometric determination of reaction rates, but definite increases in absorbance were observed. No reaction was detected in the absence of enzyme or NAD. The reverse reaction, the reduction of DDCHO, was similarly observed with the substitution of reduced NAD for the oxidized form and by following the decrease in absorbance at 340 nm. Glycerolformal in the absence of DDOH also catalyzes the reduction of NAD, but the reverse reaction was not observed for glycerolformal.

Direct chemical evidence for the enzymatic oxidation of DDOH to DDCHO was obtained by formation of the *p*-nitrophenylhydrazone derivative. An incubation mixture was prepared containing 0.016*M* sodium pyrophosphate, *p*H 8.8, 0.008*M* NAD, and 2 mg of DDOH in a final volume of 6 ml. Crystalline liver alcohol dehydrogenase (2 mg) was added, and the mixture was incubated at 37°C for 30 minutes; then 0.01*M* *p*-nitrophenylhydrazine (0.5 ml) was added, and the mixture was shaken. A chloroform extract of the mixture was prepared and evaporated to a minimum volume. Portions of the extract were chromatographed on silica-gel plates with a mixture of benzene and petroleum ether (75:25) or benzene and ethyl acetate (95:5). The extract yielded a spot on the chromatograms whose R_F values, 0.28 and 0.12 for the respective solvent systems, corresponded to those of the authentic *p*-nitrophenylhydrazone derivative of DDCHO.

The identity of the derivative obtained from the enzymatic incubation was established by low-resolution, electron-impact mass spectrometry at 20 ev. High resolution measurements (4)

confirmed the elemental composition of the ion fragments observed in the low resolution spectrum. An authentic sample of the *p*-nitrophenylhydrazone of DDCHO shows a prominent molecular ion at *m/e* (mass to charge) 399 and the base peak at 261. Other prominent ions of interest were *m/e* 249, 235, 226, 199, 200, 125, 122, and 111. The mass spectrum of the chloroform-extractable derivative was the same as that of the authentic sample of the DDCHO derivative. These findings strongly support the probability that DDCHO is a metabolite of DDT.

References and Notes

1. J. E. Peterson and W. H. Robinson, *Toxicol. Appl. Pharmacol.* 6, 321 (1964).
2. J. D. McKinney, E. L. Boozer, H. P. Hopkins, J. E. Suggs, *Experientia* 25, 897 (1969).
3. B. L. Vallee and F. L. Hoch, *Proc. Nat. Acad. Sci. U.S.* 41, 327 (1955).
4. We thank Prof. Klaus Biemann and Dr. Charles Hignite, Mass Spectral Laboratory, Massachusetts Institute of Technology, Cambridge. Massachusetts for high resolution measurements. Supported, in part by NIH research grant RR-00317 from Division of Research Facilities and Resources.

AUTHOR INDEX